IMPORTS AND POLITICS

Trade Decision-Making in Canada, 1968-1979

David R. Protheroe

Institute for Research on Public Policy/Institut de recherches politiques

Montreal
1980

ISBN 0-920380-45-X

Legal Deposit First Quarter
Bibliothèque nationale du Québec

Institute for Research on Public Policy/Institut de recherches politiques
2149 MacKay
Montréal, Québec H3G 2J2

Typesetting by Judy Bradley

Printing by Tri-Graphic Printing (Ottawa) Ltd.

Founded in 1972, the INSTITUTE FOR RESEARCH ON PUBLIC POLICY is a national organization whose independence and autonomy are ensured by the revenues of an endowment fund which is supported by the federal and provincial governments and by the private sector. In addition, the institute receives grants and contracts from governments, corporations, and foundations to carry out specific research projects.

The *raison d'être* of the institute is threefold:

— To act as a catalyst within the national community by helping to facilitate informed public debate on issues of major public interest

— To stimulate participation by all segments of the national community in the process that leads to public policy making

— To find practical solutions to important public policy problems, thus aiding in the development of sound public policies.

The institute is governed by a Board of Directors, which is its decision-making body, and a Council of Trustees, which advises the board on matters related to the research direction of the institute. Day to day administration of the institute's policies, programmes, and staff is the responsibility of the president.

The institute operates in a decentralized way, employing researchers located across Canada. This ensures that research undertaken will include contributions from all regions of the country.

Wherever possible, the institute will try to promote public understanding of, and discussion on, issues of national importance, whether they be controversial or not. The institute will publish its research findings with clarity and impartiality. It is not the function of the institute to control or influence the conduct of particular research or the conclusions reached thereby. Conclusions or recommendations in IRPP publications are solely those of the author, and should not be attributed to the Board of Directors, Council of Trustees, or contributors to the institute.

The president bears final responsibility for the decision to publish a manuscript under an IRPP imprint. In reaching this decision, he is advised on the accuracy and objectivity of a manuscript by both internal IRPP staff and outside reviewers. Publication of a manuscript signifies that it is deemed to be a competent treatment of a subject worthy of public consideration.

Publications of the institute are published in the language of the author, along with an executive summary in both of Canada's official languages.

THE MEMBERS OF THE INSTITUTE

PREFACE

One of the preeminent facts of Canadian life is the size of the foreign trade component of our Gross National Product. Despite our relatively high (but declining) tariffs on many manufactured goods, Canada is an open economy. Decisions by the federal government about tariff and non-tariff barriers to trade have an important influence on the country's industrial structure, balance of payments and the prices consumers pay for a wide variety of goods.

This study, in the words of the author, provides "a descriptive analysis of how trade decisions were made by the Trudeau government in the years 1968 to 1979." Protheroe's analysis makes use of a number of "models" of government decision making. He also gives an exceptionally clear picture of both the environment in which trade decisions are made and of the institutions and actors who make them. Protheroe, like Presthus and others, finds that most federal decisions about commercial policy can be "explained" by a process of "elite accommodation." In such cases, a co-operative problem-solving approach is adopted despite significant differences in objectives among the actors.

I believe this study will be of interest to politicians, public servants, academics and students of decision making. The improvement of public policy requires that we examine not only what was done, but also how it was done, and why it was done.

Michael J.L. Kirby
President
January, 1980

PRÉFACE

Un des faits saillants de la vie canadienne, c'est l'importance du composant commerce extérieur dans notre produit national brut. En dépit de nos tarifs relativement élevés (quoique décroissants) sur un grand nombre d'articles fabriqués, notre économie est ouverte. Les décisions du gouvernement fédéral relativement aux barrières tarifaires et non tarifaires au commerce ont une influence sensible sur la structure industrielle du pays, la balance des paiements et les prix à la consommation d'un large éventail de produits.

Dans les mots de l'auteur, cette étude présente une "analyse descriptive de la façon dont les décisions commerciales étaient prises par le gouvernement Trudeau au cours de la période de 1968 à 1979." L'analyse de Protheroe recourt à un certain nombre de "modèles" des processus décisionnels du gouvernement. Elle évoque de façon exceptionnellement claire à la fois le climat dans lequel se prennent les décisions commerciales et les institutions et les protagonistes qui les prennent. Comme Presthus et d'autres, Protheroe arrive à la conclusion que la plupart des décisions fédérales en matière de politique commerciale trouvent leur "explication" dans le processus d'"accommodements entre élites." Dans ces cas, les protagonistes, malgré la divergence marquée de leurs objectifs, choisissent de collaborer entre eux à la solution des problèmes.

Je crois que cette étude intéressera les hommes politiques, les fonctionnaires, les universitaires et les analystes du processus décisionnel. Pour améliorer la politique publique, nous devons examiner non seulement ce qui s'est fait dans le passé, mais aussi le comment et le pourquoi de ce qui s'est fait.

Michael J.L. Kirby
Président
Janvier 1980

TABLE OF CONTENTS

ACKNOWLEDGEMENTS

Confidential interviews produced most of the information for this study. Unhappily, this research technique prevents the author from thanking many individuals by name for their patience and candour during interviews. Several among them were especially generous of their time in reading drafts and offering valuable comments; this kindness is deeply appreciated.

It is permissible, however, to express special thanks to Professor Peyton V. Lyon of Carleton University's School of International Affairs. By turns a spark of ideas, giver of encouragement and prompter of acceptable prose, it is difficult to imagine a more helpful supervisor for the M.A. thesis which formed the basis for this study.

The manuscript was edited by Bill Stanbury of IRPP who also provided assistance on substantive matters. As is the case with all IRPP publications, the statements and judgments contained in this document are the responsibility of the author alone.

David Protheroe
Ottawa
January, 1980

THE AUTHOR

Born in Edmonton, Alberta, David R. Protheroe studied at Strasbourg and Carleton Universities, completing his M.A. in International Affairs in early 1979. This volume is based largely on his Master's Thesis. Since finishing his studies, Mr. Protheroe has joined the Public Service of Canada.

EXECUTIVE SUMMARY

This study of federal government trade decision-making focuses on the traditional concerns of commercial policy - tariffs and non-tariff barriers to imports. Some increasingly important issues in Canada's trade policy, such as export financing and commodity price stabilization, are given only cursory attention here.

In contrast to many IRPP studies, this one does not specifically aspire to reform the trade decision-making process in Canada. Its more restricted purpose is to provide, as objectively as possible, a descriptive analysis of how trade decisions were made by the Trudeau government in the years 1968 to 1979. This empirical orientation is reflected in our testing of several well-known decision models against information gathered in interviews with politicians, public servants and private citizens.

These decision models are outlined in Chapter One. Political scientists, historians and economists have explained government decisions by using, explicitly or implicitly, models of two broad types: "macro" models based on assessments of the broad balance of forces in society and the international environment, and "micro" models focussing, in addition to the preceding, on the interactions of protagonists within the institutions of government, particularly in the Cabinet and senior bureaucracy. The "micro" approach is the more comprehensive; this research accordingly tries to maximize its use. Unfortunately, it is also more demanding of information, and sometimes we revert to the "macro" approach when information about intra-governmental behaviour is scarce.

An example of the "macro" approach to explaining trade decisions is provided by Professor Ingo Walter's "politico-economic decision model," which sees trade outputs as the result of competing pressures from "protection-biased" and liberal"trade-biased" groups, balanced against government objectives and prevailing economic and social conditions. As far as it goes, Walter's approach is a useful way of finding plausible explanations of trade decisions. Greater certainty of explanations, however, can be expected from the micro, "look inside government" approach. Operationally, this entails identifying the individual or organizational actors in the government structure involved in decisions; tracing their views, power resources and actions; assessing the interactions between them; and searching for patterns.

Five decision models which make generalizations about such patterns of behaviour are used in this study. Two are "rational" models: decision-makers are presumed to share values and objectives and to choose among options in a cost-benefit fashion with a view to maximizing their shared goals. Sometimes these goals are substantive in character (related to the subject matter at hand). For example, tariffs might be lowered if there is a consensus that inflation-fighting is the prime thrust of macro-economic policy. This circumstance we dub the "rational-substantive" model. In other situations, the shared aim may be political - perhaps a shared desire to satisfy an interest group or region - hence the term "rational-political" model.

A special case of the rational models is called the "individual model." Its variants emphasize respectively individual leaders and individual organizations. It assumes that the leader or organization is able to impose his/its views or act in such isolation as to render analysis of interaction with other actors irrelevant in explaining outcomes. Like the first two models, this one asserts rational (political or substantive) ends-means calculation by the leader or organization.

Two other models assume disparate objectives among decision-makers. The "governmental politics" model sees decisions arising, not from rational calculation, but out of hard bargaining between actors representing various interests, such as consumers, employers, labour and regions. A final model, termed the "accommodation" model, also posits differing preferences among the key decision-makers, but in this instance the dynamic of choice is seen to be cooperative problem-solving rather than conflictual bargaining. Protagonists are not adamant about their varying aims and consciously seek decisions which reflect a mix of priorities and interests.

A major purpose of this study is to determine the frequency with which the decision models "appeared" in the record of trade decisions between 1968 and 1979 and to determine those circumstances which encourage the processes described by each. These aims are pursued by comparative analysis of several "sub--systems" of the trade policy field, namely, unilateral tariff changes, quantitative restrictions, the anti-dumping and coutervailing duty process, and the Multilateral Trade Negotiations.

Chapter Two consists of a description of the broad environment of trade decision-makers: the historical and contemporary goals and tactics of commercial policy; constraints on the freedom of action of Canadian governments; and the domestic and international legislation against which specific decisions were made in the 1968-79 period. It is intended primarily as background for the uninitiated. Readers reasonably familiar with Canadian trade policy and its legal dimensions might safely skip this chapter.

The three chapters following analyze the roles of private groups and government institutions participating in the process of making trade policy. Chapter Three describes the views and behaviour of interest groups, provincial governments and parliamentarians as initiators of decision-making sequences. After surveying the attitudes of associations in the agricultural, natural resources and manufacturing sectors, along with the interests of labour, consumers, exporters and importers, questions are posed about the relative strengths of protectionist and trade liberalization forces in Canada. The former are found to possess the stronger potential for influence in terms of their electoral and financial leverage. They do not always prevail, however, because increases in protection are often inhibited by the GATT rules, government anxiety about retaliation from trade partners, the tenor of federal macro-economic policy, the offsetting power of many liberal interest groups and regions, as well as the process of compromise and accommodation occurring within the federal government. Certain circumstances appear to favour the relative success of protectionist and liberal lobbyists. Recessions and isolated one-product cases benefit protectionists. In contrast, liberal interests are more likely to prevail in periods of economic expansion and in negotiating contexts where numerous products are on the table and where the bargain struck between countries can reflect a balance of export opportunities and concessions for each. As for provincial governments, their role as conduits for the communication of regional interests became markedly more forceful over the Trudeau years. Members of Parliament, especially Government caucus members, played a similar and not insignificant role.

Chapter Four examines the "specialized review mechanisms": the Anti- Dumping Tribunal, Tariff Board and Textile and Clothing Board. Charged with giving independent review to disputed questions prior to a Cabinet decision, these institutions were formally set up to meet Canada's obligations under the GATT. Tacitly, they provide a means of defusing politically-sensitive issues for ministers under fire from competing and vocal interests. The Textile and Clothing Board, however, has enjoyed markedly less success in defusing controversy than the others, largely due to less precise international and domestic legislation governing its mandate.

Chapter Five considers the roles of public servants in departments and central agencies (plus the Economic Council and Prime Minister's Office) as reviewers of options and advisors to ministers. Compared with many areas of public affairs, trade policy saw participation by a large number of departments and agencies in the 1968-79 period. This doubtless reflected the diverse regional, macro-economic, socio-economic and foreign policy implications of trade decisions, as well as the Trudeau government's partiality to collegial decision-making. These departments and central agencies, and in some cases their sub-units, differed substantially in their attitutes toward trade issues. Three main factors account for these value cleavages (ranging, roughly, on a scale from "protectionist" to "liberal"). Each organization's function in the total government system is a major determinant. Portfolios are classified into "vertical constituency" departments, which operate programs benefiting some social group or economic sector and which tend to represent their views on trade issues; "horizontal coordinative" portfolios that cut across several responsibilities and have minimal links to particular interests; and "administrative coordinative" departments, which are largely concerned with implementing decisions. Clear examples are, respectively, the Departments of Fisheries, Finance and National Revenue. Of perhaps equal weight in determining departmental stances are the individual interests and views of incumbent ministers, deputy heads and assistant deputy ministers. A final factor is the role of departmental sub-units in the trade policy division of labour. Units whose major function is to negotiate away foreign barriers to Canadian exports, for instance, tend to favour trade liberalization more than do units administering government assistance to domestic constituencies.

Chapter Six presents a case study using the various decision models as analytical tools. It deals with the imposition of quantitative controls on "low-cost" textile and clothing imports. The evidence gathered suggests that in the bulk of the more than 50 instances of quantitave restraints applied from the introduction of a new Textile Policy in 1971 to the end of 1976, decisions were arrived at in close conformity to the "governmental politics" model. Decisions were reached through a process of bargaining and compromise between dissenting protagonists at the bureaucratic and Cabinet levels of the federal government. The typical process, however, was in a few cases replaced by rapid emergence of a political consensus at the level of ministers, such that the "rational-political" model became the most satisfying explanation. A major example of the latter process was observed in the November 1976 decision to invoke global quotas on virtually all clothing imports. A growing politicization of the clothing issue in the summer and fall of 1976, culminating in the shock to federal politicians of the Parti Québécois' victory at the polls on November 15, caused ministers to set aside many of those regional and departmental considerations which had guided their behaviour on such issues in the past. This unusual consensus resulted in a more extreme (protectionist) decision on clothing imports than might otherwise have been the case.

Chapter Seven summarizes the general features of trade decision-making in the Trudeau government - the variable patterns of participation by departments and ministers in the decision process, cleavages between political and bureaucratic actors, and the typical ways of resolving differences. While conflicts over the socio-economic, regional and political dimensions of trade issues were endemic, in most cases certain norms of behaviour eased their resolution (if not full agreement). These included acceptance of the "lead department" principle, which sets limits to opposition from non- responsible departments, and the informal hierarchy of ministers and portfolios.

A special issue addressed is the relative weight of senior bureacrats and ministers in making trade decisions. This study does not directly support the contentions of many observers of Canadian politics who have argued that real power has shifted in recent years away from ministers toward unelected public servants. Our conclusion is that ministers very definitely did, for good or ill, make trade decisions in the Trudeau government. This was done either through personal action or because, although in some instances their personal involvement may have been minimal, their advisors successfully anticipated the political and regional interests of ministers in their recommendations. But scant comfort may be drawn from this conclusion by the proponents of "ministerial responsibility and political control," for trade policy may well not be, in this sense, typical of government decisions in general. On the whole, trade issues are not technically or scientifically beyond the grasp of non-specialist ministers. Their political sensitivity, moreover, renders them instinctively understandable to politicians.

Chapters Eight and Nine present a comparative analysis of decision-making in several trade policy "subsystems." Unilateral tariff decisions (that is, unrelated to GATT rounds) were found to correspond largely to the accommodation or problem-solving model. This was so principally because of the Department of Finance's reluctance to allow tariffs to be subjected to interdepartmental bargaining and because Finance is a horizontal department without evident constituency linkages and as such was wont to take some account of all sides to the issues. One tariff issue, however, proved a departure from the accommodation norm. External pressure saw Finance drawn into the unaccustomed experience of chairing an interdepartmental committee on the General Preferential Tariff. In this case the bargaining or "governmental politics" model best describes the committee's deliberations.

The Multilateral Trade Negotiations between 1973 and 1979, the second subsystem considered, seemed explainable in terms of two decision models. In the early, goal-setting phase of the negotiations, a wide consensus existed in Ottawa and the provinces with respect to Canadian objectives, especially reduction of foreign trade barriers which inhibit upgrading of Canadian resources prior to export. The "rational-substantive" model thus marked this easiest stage of the MTN. Consensus was more elusive when it came to specifying Canadian concessions to be exchanged for foreign reductions in tariffs and non-tariff barriers. The Ottawa principals frequently were in disagreement on these issues and on tactical questions. All, moreover, were aware of conflicting pressures from interest groups and the provinces. The decision dynamic did not, however, become one of conflictual bargaining. Rather, several reasons encouraged the problem-solving approach described by the "accommodation" model: most principals were not adamant about the bulk of issues and sought a balance of concessions from and gains to Canada. The "vertical constituency" departments, moreover, recognized the necessity of finding solutions representing a mix of the interests of their export-oriented and import-vulnerable sectors (e.g., Western and Eastern agriculture). These encouraged flexibility among decision-makers, as did probably the experience of negotiating with stronger trade partners. A third policy subsystem, the Anti-Dumping and countervailing duties process, exhibited most clearly the traits of the "rational-substantive" model. This subsystem has been largely "depoliticized" through quite precise statutes, administrative regulations, and international law.

The study ends with a discussion of the frequency of and the conditions requisite to the functioning of the different decision models. In contrast to much of the thrust of recent political science writings, conflict theories of the federal political system receive less support than the so-called "elite accommodation" approach to Canadian politics. The accommodation or problem-solving model was predominant among the cases we considered. But it was not always applicable. Issues of unusual and multi-faceted controversy - such as imports from low wage countries - appeared to induce hardline positions among a variety of principals based on an assortment of motives. The remaining models appeared to describe trade decisions much less frequently, but they sometimes concerned very important decisions (e.g., the "rational-political" model in the 1976 clothing quotas case).

ABRÉGÉ

Cette étude du processus décisionnel du le gouvernement fédéral dans le secteur commercial est axée sur les problèmes traditionnels de toute politique commerciale: les barrières tarifaires et non tarifaires aux importations. Certaines questions qui ont une importance croissante dans la politique commerciale canadienne, telles que le financement des exportations et la stabilisation des prix des marchandises, ne sont abordées ici que superficiellement.

Contrairement à nombre d'études entreprises par l'IRPP, celle-ci ne vise pas particulièrement à réformer le processus décisionnel du gouvernement en l'espèce. Son objet se limite à présenter, le plus objectivement possible, une analyse descriptive de la façon dont les décisions commerciales étaient prises sous le governement Trudeau au cours de la période de 1968 à 1979. Cette orientation empirique est reflétée dans notre évaluation de plusieurs modèles courants de prises de décisions que nous avons mis en regard des informations recueillies au cours d'entrevues avec des hommes politiques, des fonctionnaires et des citoyens.

Ces modèles de décisions sont décrits au chapitre un. Experts en science politique, historiens et économistes ont expliqué les décisions du gouvernement en utilisant, explicitement et implicitement, des modèles de deux types généraux: les "macromodèles", fondés sur l'évaluation de l'équilibre général des forces dans la société et sur le plan international, et les "micromodèles," qui visent, par surcroît, l'interaction des protagonistes dans les institutions gouvernementales, tout particulierement au sein du Cabinet et des cadres supérieurs de l'Administration. La méthode du micromodèle est de portée plus étendue que l'autre. Nous avons donc tenté d'y recourir le plus largement possible dans cette étude. Malheureusement, la méthode exige aussi de plus amples données de base, de sorte que nous avons à l'occasion adopté l'approche du macromodèle lorsque nous manquious de renseignements sur l'interaction des forces au sein du gouvernement.

Un exemple de "macrométhode" visant à expliquer les décisions commerciales nous est donné par le professeur Ingo Walter dans son "modèle de décisions politico-économique". Ce modèle interprète les décisions commerciales comme le résultat de pressions concurrentielles entre les groupes qui penchent pour le protectionnisme et, d'autre part, les groupes qui favorisent le libre échange, ces pressions étant équilibrées par les objectifs gouvernementaux et la conjoncture économique et sociale. L'approche de Walter constitue un moyen utile de trouver des explications plausibles à la nature des décisions commerciales. La "microméthode" d'observation "interne" du gouvernement offre cependant beaucoup plus d'assurance quant à la validité des explications qu'on peut en tirer. Concrètement, cela suppose qu'on détermine les "acteurs" -- personnes ou organismes -- qui sont impliqués dans les décisions au sein du gouvernement en définissant leurs points de vue, l'étendue de leurs pouvoirs et leur actions, en déterminant leur interaction et en recherchant dans cette analyse les éléments qui se prêtent à la généralisation.

Dans cette étude, cinq modèles de décisions sont utilisés, modèles qui font des généralisations sur ces types de comportement. Deux d'entre eux sont des modèles dits "rationnels": on assume que les décisionnaires partagent les mêmes idées, les mêmes objectifs et choisissent parmi les options en présence du point du vue coût-bénéfices pour maximiser leurs objectifs communs. Quelquefois ces buts ont un caractère substantif (c'est-à-dire qu'ils sont reliés à une question déterminée). Par exemple, les tarifs peuvent être abaissés s'il existe un consensus par lequel la lutte contre l'inflation est le premier objectif de la politique macro-économique. Dans ce contexte, nous qualifions le modèle de "substantivo-rationnel". Dans d'autres cas, le but partagé peut être politique -- par exemple le désir partagé de satisfaire un groupe d'intérêt ou une région --, d'où le terme modèle "politico-rationnel".

Un type spécial de modèle rationnel est appelé "modèle individuel". Ses variantes mettent respectivement l'accent sur les dirigeants et les organisations individuels. Ce modèle suppose que le dirigeant ou l'organisation est capable d'imposer ses vues ou d'agir isolément de façon à rendre l'analyse de son interaction avec les autres acteurs totalement inutilisable pour expliquer les mesures prises. Comme les deux premiers modèles, ce dernier fait valoir le calcul rationnel (politique ou substantif) des moyens et fins par le dirigeant ou l'organisme.

Les deux autres modèles reposent sur l'hypothèse que les décisionnaires recherchent des buts différents. Le modèle de "politiques gouvernementales" voit les décisions émanant, non pas d'un calcul rationnel, mais de négociations difficiles entre les acteurs représentant différents intérêts, tels que les consommateurs, employeurs, travailleurs et régions. Un dernier modèle, intitulé le modèle d'"accommodement", définit également les différentes préférences parmi les principaux responsables des décisions, mais, dans ce cas, la dynamique du choix se manifeste dans la recherche concertée des solutions, non dans les situations conflictuelles de la négociation. Les protagonistes ne se montrent pas intransigeants quant aux buts différents qu'ils visent: ils cherchent sciemment à réaliser des décisions qui traduisent une composition de priorités et d'intérêts.

L'un des principaux objets de cette étude est de déterminer la fréquence avec laquelle les modèles de décisions "se manisfestent" dans les décisions commerciales consignées entre 1968 et 1979 et de déterminer les circonstances qui appuient la méthode décrite par chacun d'eux. Ces buts sont poursuivis au moyen d'une analyse comparative de plusieurs "sous-systèmes" du domaine de la politique commerciale, plus précisément, les modifications unilatérales de tarifs, les restrictions quantitatives, les mesures antidumping et les taxes compensatrices, ainsi que les négociations commerciales multilatérales.

Le chapitre deux consiste en une description de l'environnement général dans lequel évoluent les décisionnaires en matière commerciale: les buts et les tactiques historiques et contemporaines de la politique commerciales; les entraves à la liberté d'action des gouvernements canadiens et les lois nationales et internationales en vertu desquelles les décisions particulières ont été prises au cours de la période 1968-1979. Ce chapitre a été conçu à l'intention des non-initiés. Les lecteurs qui connaissent assez bien la politique commerciale canadienne et ses aspects juridiques peuvent en toute confiance sauter ce chapitre.

Les troise chapitres suivants analysent les rôles des groupes privés et des institutions gouvernementales qui participent au processus d'élaboration de la politique commerciale. Le chapitre trois décrit les points de vue et le comportement des groups d'intérêt, des gouvernements provinciaux et des parlementaires dans le déroulement du processus décisionnel qu'ils mettent en branle. Après avoir étudié les attitudes des associations dans les secteurs de l'agriculture, de l'industrie et des ressources naturelles, ainsi que les intérêts des travailleurs, consommateurs, exportateurs et importateurs, la question se pose de savoir quelle est la puissance relative au Canada des forces protectionnistes d'une part et, d'autre part, des mouvements de libéralisation du commerce. Il se trouve que les premiers possèdent le plus fort potentiel d'influence sur le plan électoral et financier. Cependant, ils ne l'emportent pas toujours, car l'intensification du protectionnisme est souvent freinée par les régles du GATT, l'inquiétude du gouvernement quant aux représailles éventuelles de ses partenaires commerciaux, le sens général de la politique macroéconomique fédérale, le contrepoids de beaucoup de groupes d'intérêt et de régions favorables à la libéralisation du commerce, ainsi que le processus de compromis et d'accommodement existant au sein du gouvernement fédéral. Certaines circonstances semblent favoriser le succès relatif des lobbies du protectionnisme et du libéralisme. Les récessions et les cas isolés où un seul produit est en jeu servent les intérêts des protectionnistes. A l'inverse, les intérêts libéraux l'emporteront vraisemblablement dans les périodes d'expansion économique et dans les circonstances où de nombreux produits font l'objet de négociations entre les pays et où les accords réalisés entre eux traduisent un équilibre entre les concessions que chacun consent en retour des avantages à l'exportation qui lui sont consentis. En ce qui concerne les gouvernments provinciaux, leur rôle d'interprètes des intérêts régionaux s'est sensiblement affirmé pendant les années Trudeau. Les membres du Parlement et, en particulier, les membres du caucus gouvernemental ont joué un rôle similaire et non moins important.

Le chapitre quatre examine les "organises spécialisés de révision": le Tribunal antidumping, la Commission du tarif et la Commission du textile et du vêtement. Chargées de réviser indépendamment les questions litigieuses avant les décisions du Cabinet, ces institutions furent établies à l'origine pour répondre aux obligations du Canada vis-à-vis du GATT. Tacitement, elles offrent aux ministres assaillis par des intérêts opposés et militants un moyen de trouver des solutions politiques à certaines questions épineuses. La Commission du textiles et du vêtement a eu cependant beaucoup moins de succès que les autres dans la résolution des controverses, en grande partie à cause des législations internationale et nationale moins précises qui régissaient son mandat.

Le chapitre cinq examine le rôle des fonctionnaires au sein des ministères et des institutions centrales (plus le Conseil économique et le Cabinet du Premier ministre) en tant que réviseurs d'options et conseillers des ministres. Comparé à beaucoup de domaines des affaires publiques, un nombre assez important de ministères et d'institutions gouvernementales ont participé à la politique commerciale au cours de la période 1968-1979. Cela reflète sans doute les répercussions des décisions commerciales au niveau régional, macro-économique et socio-économique et dans le domaine de la politique étrangère, ainsi que la prédilection du gouvernement Trudeau pour les prises de décision collégiales. Ces ministères et institutions centrales et, dans certains cas, leurs sous-divisions, ont abordé de façons sensiblement différentes les questions commerciales. Trois facteurs principaux sont à l'origine de ces divergences de vues (variant, *grosso modo*, du "protectionnisme" au "libre échange"). Chaque fonction au sein de l'organisation dans l'ensemble du système gouvernemental constitue un facteur déterminant. Les portefeuilles sont classés en départements par "découpage vertical," appliquent des programmes profitant à un groupe social ou à un secteur économique et tendent à présenter leurs points de vue sur les solutions commerciales; les portefeuilles à "coordination horizontale", qui recoupent plusieurs types de responsabilités et ont un minimum de rapports avec des intérêts particuliers; et les départements à "coordination administrative", qui ont la responsabilité principale d'appliquer les décisions. Pour donner des exemples clairs, on peut citer les ministères des Pêcheries, des Finances et du Revenu national. Les intérêts individuels et les opinions des ministres en place, des sous-ministres et des sous-ministres adjoints contribuent peut-être dans la même mesure à déterminer les positions des départements en l'espèce. Le dernier facteur, c'est le rôle des sous-divisions au sein des départements dans la division du travail de la politique commerciale. Ces unités qui émanent des ministères et dont la mission principale est d'abattre par voie de négociation les barrières douanières qui entravent les exportations canadiennes tendent à favoriser la libéralisation du commerce plus que ne le font les unités administratives chargées de l'assistance gouvernementale aux différents groupes nationaux.

Le chapitre six présente une étude de cas au moyen des différents modèles de décisions qui constituent des outils d'analyse. Il traite de l'imposition des contrôles quantitatifs sur les textiles "à bas prix" et les importations de vêtements. Les informations qui en sont tirées montrent que, d'une manière générals, dans le cas des 50 mesures de restrictions quantitatives appliquées en vertu de l'introduction, en 1971, de la nouvelle politique des textiles, qui est restée en vigueur jusqu'à la fin de 1976, les décisions ont été prises en étroite conformité avec le modèle de "politiques gouvernementales". On est parvenu à ces décisions par un processus de négociations et de compromis entre les différents protagonistes au niveau administratif et ministériel du gouvernement fédéral. Cependant le processus typique, dans quelques cas seulement, a été remplacé par la formation rapide d'un consensus politique au niveau ministériel, de sorte que le modèle "politico-rationnel" est devenu l'explication la plus satisfaisante. On a pu observer un exemple typique de ce processus en novembre 1976 lorsque la décision a été prise d'appliquer des quotas globaux à toutes les importations de vêtements. Au cours de l'été et de l'automne 1976, la question a acquis un caractère politique de plus en plus marqué, pour atteindre son point culminant lors de la victoire du Parti Québécois aux élections du 15 novembre: ébranlés par cet événement, les ministres du Cabinet fédéral abandonnènent bon nombre des considérations d'ordre régional et ministériel qui, par le passé, avaient guidé leur comportement politique dans les questions de cette nature. Ce consensus inusité entraîna une prise de position protectionniste encore plus rigoureuse quant aux importations de vêtements que ce n'eût sans doute été le cas dans d'autres circonstances.

Le chapitre sept résume les principales caractérisques du processus décisionnel en matière commerciale sous le gouvernement Trudeau: les modes variables de la participation des départements et des ministres dans le processus, les scissions entre protagonistes politiques et administratifs et les façons typiques de concilier leurs positions respectives. Bien que les conflits à propos des différents aspects socio-économiques, régionaux et politiques des problèmes fussent chose commune, dans la plupart des cas, cependant, certaines normes de comportement en ont facilité la résolution (sinon entraîné le parfait accord des parties). L'une de ces normes, c'était la reconnaissance de principe d'un lead department, ou ministère-clé, qui tempère l'opposition des ministères non responsables, et l'acceptation d'une hiérarchie de facto des ministres et des portefeuilles.

Un autre sujet particulier est abordé: le poids relatif des cadres supérieurs et des ministres dans la prises des décisions commerciales. Cette étude ne soutient pas directement la thèse de beaucoup d'observateurs de la scène politique canadiennes selon lesquels le pouvoir réel, au cours des dernières années, est passé des ministres à des fonctionnaires non élus. Notre conclusion est que les ministres, sous le gouvernement Trudeau, ont effectivement pris les décisions (bonnes ou mauvaises) en matière de commerce, soit qu'ils soient intervenus personnellement de leur chef, soit que leurs conseillers aient correctement prévu dans leurs recommandations les intérêts politiques et régionaux des ministres, bien que, dans certains cas, le rôle de ces conseillers puisse avoir été négligeable. Mais les tenants de la "responsabilité ministérielle" et du "droit de regard politique" trouveront peu de réconfort dans cette conclusion, car il se peut fort bien que, sous cet aspect, la politique commerciale ne soit pas caractéristique des décisions gouvernementales dans leur ensemble. Dans l'ensemble, les questions commerciales ne sont pas, techniquement ou scientifiquement, inaccessibles aux ministres non spécialistes en la matière. La nature politiquement délicate de ces questions fait que les hommes politiques les comprennent d'instinct.

Les chapitres huit et neuf présentent une analyse comparative des prises de décision au sein de plusieurs "sous-systèmes" de politique commerciale. Les décisions unilatérales sur les tarifs (c'est à dire non liés aux accords du GATT) se sont révélées largement apparentées au modèle d'accommodement ou de résolution des problèmes. La raison en est principalement que le ministère des Finances répugne à faire des tarifs un sujet de négociations interministérielles, que c'est un ministère "horizontal", sans liens évidents avec les ministères verticalement constitués, et que, de ce fait, il tend habituellement à prendre en considération tous les aspects du problème. Dans un cas en particulier, cependant, il y a eu dérogation à la norme d'accommodement. Des pressions exercées de l'extérieur ont amené les Finances a se lancer dans l'expérience inhabituelle de présider un comité interdépartemental sur le Tarif préférentiel général. Dans ce cas, le modèle de négociations ou de "politiques gouvernementales" décrit le mieux les délibérations du comité.

Les négociations commerciales multilatérales entre 1973 et 1979, le second sous-système envisagé, semblaient explicables selon deux modèles de décisions. Dans la phase initiale des négociations, consacrée à la définition des objectifs, un large consensus existait entre Ottawa et les provinces quant aux objectifs canadiens, en particulier la réduction des barrières douanières étrangères qui entravent la mise en valeur des ressources canadiennes avant exportation. Le modèle "substantivo-rationnel" a donc marqué cette étape plus facile des négociations. Le consensus a été plus élusif le moment venu de définir les concessions que le Canada consentinait aux pays étrangers en contrepartie des réductions de leurs barrières tarifaires et non tarifaires. Les représentants d'Ottawa étaient fréquemment en désaccord sur ces questions et sur les questions d'ordre tactique. En outre, ils étaient tous conscients des pressions contradictoires qu'exerçaient les groupes d'intérêt et les provinces. La dynamique des décisions n'a cependant pas dégénéré en négociations conflictuelles. En fait, plusieurs facteurs ont favorisé la méthode de résolution des problèmes décrite dans le modèle "accommodement". La plupart des protagonistes ont témoigné de souplesse sur le gros des questions en jeu et ont recherché un équilibre entre les concessions que le Canada pourrait faire et les avantages qui lui reviendraient. Les départements à "découpage vertical" ont d'ailleurs reconnu la nécessité de trouver des solutions servant un amalgame d'intérêts de leurs secteurs tournés vers les exportations et ceux de leurs secteurs vulnérables aux importations (e.g., l'agriculture de l'Ouest et celle de l'Est). Cette attitude -- de même, vraisemblablement, que l'expérience de négocier avec des partenaires commerciaux plus puissants -- a incliné les décisionnaires à la souplesse. Un troisième sous-système de politique, la méthode antidumping et les taxes compensatrices, a mis plus clairement en évidence les caratéristiques du modèle "substantivo-nationnel". Ce sous-système a été largement "dépolitisé" par des lois tout à fait précises, par des règlements administratifs et par le droit international.

L'étude se termine par une discussion sur la fréquence des différents modèles de décisions et les conditions nécessaries a leur application. Contrairement à un tendance générale des travaux de science politique récents, les théories sur les conflits du système politique fédéral reçoivent ici moins de crédit que le rôle des "accommodements entre élites" dans la politique canadienne. Le modèle d'accommodement ou de résolution des problèmes prédominait dans la plupart des cas que nous avons étudiés, mais il n'était pas toujours applicable. Les questions inusitées et qui, par bien des côtés, prêtent à la controverse -- telles les importations des pays où les salaires sont bas -- semblent avoir amené une variété de dirigeants à raidir leurs positions pour diverses raisons. Les autres modèles ont paru décrire beaucoup moins fréquemment les décisions commerciales, mais ils s'appliquaient parfois à des décisions très importantes (par exemple, le modèle "politico-nationnel" dans le cas des quotas de 1976 sur les vêtements).

GLOSSARY OF ACRONYMS

ADM	Assistant Deputy Minister
BP	British Preference
CAC	Consumers' Association of Canada
CCA	Department of Consumer and Corporate Affairs
CEA	Canadian Export Association
CIA	Canadian Importers Association
CIDA	Canadian International Development Agency
CLC	Canadian Labour Congress
CMA	Canadian Manufacturers' Association
CTTC	Canadian Trade and Tariffs Committee
DEA	Department of External Affairs
DM	Deputy Minister
DREE	Department of Regional Economic Expansion
EEC	European Economic Community
E&I	Department and Commission of Employment and Immigration
GATT	General Agreement on Tariffs and Trade
GPT	General Preferential Tariff
GSP	Generalized System of Preferences
ICCP	Interdepartmental Committee on Commercial Policy

ICERDC	Interdepartmental Committee on Economic Relations with Developing Countries
IER	International Economic Relations Division (Department of Finance)
IPD	International Programs Division (Department of Finance)
IT&C	Department of Industry, Trade and Commerce
ITF	International Trade and Finance Branch (Department of Finance)
ITR	International Trade Relations Branch (Department of Industry, Trade and Commerce)
LCI Committee	Interdepartmental Committee on Low-Cost Imports
LDCs	Less Developed Countries
LTA	Long Term Arrangement Regarding International Trade in Cotton Textiles
MFA	Multifibre Agreement
MFN	Most Favoured Nation
MTN	Multilateral Trade Negotiations (Tokyo Round)
NIEO	New International Economic Order
OECD	Organization for Economic Cooperation and Development
PCO	Privy Council Office
PMO	Prime Minister's Office
SSEA	Secretary of State for External Affairs
TCB	Textile and Clothing Board

TNCC	Trade Negotiations Coordinating Committee
UNCTAD	United Nations Conference on Trade and Development
VER	Voluntary Export Restraint agreements

INTRODUCTION

It is not surprising that trade policy has been a subject of sustained interest on the part of Canadian academics and opinion makers for over a century. The traditional international debate between protectionists and liberals over the effects of trade decisions on employment, inflation, and resource allocation between economic sectors has of course had its Canadian version. Compared with most countries, however, trade has been more important to Canada, both economically, because of the small domestic market, and politically, because of the strong regional orientation of Canadian federalism and the fear that closer trading ties with the United States - the most obvious economic method of expanding markets for Canadian producers - would endanger national independence. This potential for conflict between economic and political values perhaps best accounts for the uncommonly great interest in trade issues over so many years by Canadians. Indeed, some of the most notable contributions of Canadians to the internationally valued literature of economics and history, including Viner, Johnson and Innis, have been made in this field.[1]

The voluminous writings on Canadian trade policy have overwhelmingly concentrated on evaluations of past decisions and scrutiny of potential trade options. The present study attempts to add in some measure to this literature by examining the federal government decision-making process on trade issues during the period of the Trudeau government, 1968 to 1979. The fundamental rationale for such a study rests on the hypothesis that structures of decision making and the interactions of actors in this process in some way affect the content of outcomes or decisions. It is also of some interest as a case study in the growing field of public policy analysis, which seeks to understand how decisions are made in political systems.

The concept of trade policy includes a broad range of activity. Basically, it refers to the management by governments of the flow of goods across borders in relation to domestic and international economic and political goals. Within this definition, however, could fall several specific activities, including the creation and dismantling of import and export barriers, export promotion programs, export financing arrangements, establishing and adapting domestic and international legal frameworks (such as the anti-dumping legislation and the General Agreement on Tariffs and Trade), international commodity agreements, and adjustment assistance measures. This study concentrates overwhelmingly on Canadian decisions, be they in a unilateral, bilateral or multilateral context, relating to tariff and non-tariff *import barriers*. Only passing attention is paid to most of the other activities.

The focus here, moreover, is on specific decisions to change or not to change individual barriers, or in the case of multilateral negotiations, groups of barriers, rather than on the more general goal-means statements and behaviours implied by the term "policy." We recognize that political scientists have had some difficulty in formulating satisfactory distinctions between the concepts of "policy" and "decisions," and we shall not attempt a resolution of that problem here. Our use of the terms is pragmatic, based on gradations of specificity where decisions are more specific (e.g., concerning individual products) than policies. Examples of "policies," as the term will be used here, include statutes which set out principles and procedures for making decisions (e.g., the *Anti-Dumping Act* and the *Textile and Clothing Board Act*), even more general statements of intent (e.g., the relevant sections of *Foreign Policy for Canadians*)[2], and often unstated traditions of behaviour (such as the reciprocity approach to trade liberalization). We will certainly not ignore these policies, but we regard them here as the background against which decisions are made, rather than the subject of our analysis per se.

Besides the accent on specific decisions, there are other limitations of scope to this study. Its time period is limited to 1968 - 1979 and thus our conclusions about decision-making processes cannot be assumed to apply to governments prior to 1968, or to the new Conservative government elected in May 1979. The Trudeau government was characterized, for example, by patterns of relations between ministers, and between political and bureaucratic levels of the executive, which were in some ways unique to it. In addition, we have based some of our conclusions on one case study, the problem of low-cost textile imports. Since this case concerns the difficulties of import competition from low-wage developing countries, it is not necessarily typical of most trade decisions. In spite of these limitations, we believe that the study reaches some broadly valid conclusions.

This study, adapted from an M.A. thesis at Carleton University's School of International Affairs, is primarily based on information collected from approximately 100 interviews conducted in Ottawa between the Summer of 1977 and the Fall of 1978 with public servants, politicians and interest group representatives. The wealth of academic and government literature on Canadian commercial policy was consulted for background and historical information, but it proved largely inadequate for a consideration of the trade decision-making process itself: hence the necessity of personal interviews. We recognize the limitations of the interview technique, but hope that our habit of checking sources against one another, as well as the fairly large number of interviews conducted, will reduce errors to a minimum.

NOTES: Introduction

1. See, in particular, Harold A. Innis, The Fur Trade in Canada: An Introduction to Canadian Economic History (New Haven: Yale University Press, 1930); Harry G. Johnson, International Trade and Economic Growth (London: G. Allen and Unwin, 1978); Jacob Viner, The Customs Union Issue (New York: Carnegie Endowment for International Peace, 1950); and Studies in the Theory of International Trade (New York: Harper and Brothers, 1937).

2. Canada, Department of External Affairs, Foreign Policy for Canadians (Ottawa: Information Canada, 1970). This document was the Trudeau government's formal statement of its international objectives.

CHAPTER ONE

DECISION MODELS AND THE ANALYSIS OF TRADE DECISIONS

The outputs of government decision-making processes have been analysed by political scientists, economists and historians at at least two levels. Perhaps the most traditional approach involves the use of models emphasizing the broad balance of forces in society. (A "model" can be regarded as an analytic tool which highlights one or several of the range of potentially important explanatory variables, thus presenting a simplified but presumably fairly accurate representation of complex reality.) These "macro" models primarily focus on inputs into the governmental process - the claims of interest groups, provinces and foreign governments - as well as environmental factors such as prevailing economic and diplomatic conditions. A second group of models regards these inputs and environments as background data and shifts attention to the interactions between government decision-makers themselves and the structures of policy- making within the complex apparatus of government. Decisions are explained by tracing in some detail the actions and arguments of actors in the governmental process - particularly in Cabinet and the senior bureaucracy - as well as the relations between these actors and organizations in the private sector.

This micro, "look inside government" modelling has the advantage of being a more comprehensive approach to the explanation of decisions than the macro approach. Its major disadvantage is that it requires more detailed information about "what happened" in particular cases, and this information is often hard to come by, especially with respect to Cabinet deliberations, around which the norm of secrecy is very strong. For these reasons, this study will employ both approaches, with some preference for micro models but sometimes settling for the less complete macro approach according to the availability of information.

A good example of the macro or input approach to explaining trade decisions is Professor Ingo Walter's article "How Trade Policy is Made: A Politico-economic Decision System."[1] Although based on American experience, it is probably the most coherent statement of how economists and historians have traditionally analysed trade decisions in Canada.

Walter's article enumerates the important factors or inputs that go into the making of any government's trade policy and presents a few hypotheses about the relative weights of these inputs. He begins with the question: what reasons may account for changes up or down or not-at-all in a country's barriers to trade? The reasons suggested are the influence of the "protection-biased sector" (most typically composed of import competing firms and unions, suppliers of these firms, and political representatives of the regions in which these groups are located), the "trade-biased sector" (generally exporting firms and farmers who wish to maximize access to foreign markets, consumer groups, importers, retailers, and manufacturers seeking low-cost imported inputs), government goals relating to some conception of the "national interest" or collective welfare, and finally, what Walter calls the "underlying economic and social conditions" which vary over time and favour either liberalization or protectionism.

Walter's model thus views trade outputs as the result of competing interests putting pressure on governments, which have goals of their own, in a given socio-economic context. He does offer several hypotheses about this process, among which the arguments that the protection-biased sector is in an inherently strong political position and that recessionary periods favour protectionists more than expansionary periods. Walter's model is capable of producing plausible explanations of trade decisions and, in the absence of more complete information about "inside government" micro relationships, these might be accepted as satisfactory explanations. But assuming that the analyst can acquire this more detailed information, it would be useful to carry the analysis a few steps further by asking the questions: What is the dynamic of the intra-governmental process? Just how do governmental actors interact between themselves and with private sector groups and foreign governments?

We are aware of five distinct micro models which have been used by Canadian political scientists (and others) to answer these questions in a variety of policy fields: a "rational-substantive" model, a "rational-political" model, a representational or "governmental politics" model, an "accommodation" model, and an "individual" model. They may be distinguished according to several characteristics. First, all except the individual model see decisions arising out of the interaction of Cabinet and bureaucratic actors. Second, the thought processes of decision-makers and the dynamic of decision-making may be rational (that is, characterized by cost-benefit calculations based on goals of the individual or, in situations of collective decision-making, shared goals and values) or non-rational. The objectives involved may be substantive (related to the subject matter at hand) or political. An example of substantive rationality would be calculation of the optimum trade instrument to achieve some shared economic value, whereas political rationality would suggest choice among trade options to achieve electoral ends. The thought processes and group dynamics are identical, though outcomes may be very different. A third point is that all models, save the "rational-substantive," are in the broad sense "political" models. For this reason the "rational-substantive" model is often viewed as an ideal. Our society tends to look down upon government decision-making processes which are perceived as "playing politics with the national (economic, military, etc.) interest."

A brief description of the sources and traits of each model follows:

(1) Rational-Substantive Model: Derived from the "pure" micro-economic theory of business decision-making, the model posits shared objectives among decision-makers in the form, say, of a clear set of industrial priorities. The costs and benefits of trade policy alternatives are calculated against these goals.

In a 1978 speech,[2] Mr. J.H. Warren, the Canadian Coordinator for the Multilateral Trade Negotiations, gave an account of the government's decision to accept the so-called Swiss tariff-cutting formula. His description, if accurate, is an excellent example of the workings of the rational-substantive model. Warren argued that, in spite of much opposition to the formula from domestic interests, like the Canadian Manufacturers' Association, Cabinet calculated the costs and benefits of various courses of action and concluded, on the basis of its major shared objective for the negotiation - better access to foreign markets for Canadian exporters - that acceptance of the formula was in the national interest. The research conducted for this study, however, indicates that there are very few real world examples of this model in the trade field, primarily because on trade issues shared values or objectives among decision-makers are quite rare. Consequently, the "national trade interest" is seldom as evident as it is for example in the field of military security. Probably the issue where this model has most frequent explanatory value is the anti-dumping system where, by international agreement, the decision-making process has been largely depoliticized and reduced to a series of technical steps based on fairly precise statutes and regulations.

(2) Rational-Political Model: This model is the classic model of political science, and is often termed simply, the "political" model. The collective Cabinet has shared (political) objectives and calculates its trade response in rational, cost-benefit fashion. In one version, R.E. Caves analysed the Canadian tariff structure against what he called the "adding machine model" which "assumes that the government chooses policies to maximize the chances of being returned to power," and found that the model had some predictive power.[3]

In what circumstances does this model seem of greatest explanatory value? By tradition, Canadian Cabinet ministers possess three major functional roles - as heads of departments, regional representatives and members of political parties. This research indicates that it is on trade issues where the party interests of ministers are most evident that objectives are likely to be sufficiently common to allow this political cost-benefit calculation. Their departmental and regional roles are on trade issues (though not necessarily other fields of public policy) likely more often than not to be sources of division. While it seems true to say that there arise relatively few trade issues where the party interest is so evident as to overcome the differing regional and departmental interests of ministers, the model does characterize a small number of highly important decisions. An example was the imposition of massive doses of protection for the clothing industry, largely located in Quebec, in the wake of the Parti Quebecois electoral victory of November 1976 (see Chapter 6).

(3) Governmental Politics or Representational Model: With this model we move away from the realm of consensus and cost-benefit calculation among decision-makers. Popularized in political science by Graham Allison,[4] and given a semi-polemical application to Canada in a recent article by senior civil servant A.W. Johnson,[5] it starts with the observation of conflicting objectives among ministers and bureaucrats. Decisions arise, not out of rational calculation, but out of bargaining and negotiation between actors which "represent" various dimensions of an issue, such as the producer, consumer, regional and other interests involved in trade problems. The result of this bargaining and conflict may be compromise, ambiguous outcomes, stalemate, or victory for one side.

It should be stressed that this bargaining may take place at the levels of both senior officials and ministers of Cabinet. There are differences between the tone of bargaining at the two levels. At the bureaucratic level, bargaining is likely to be based primarily on departmental definitions of their "missions" and links with domestic and international client groups. Being politically astute, however, these officials will to some extent tailor their advice to the political (regional and party) interests of their ministers. At the Cabinet level, bargaining may reflect as much each minister's regional/constituency roles, and varying conceptions of the party interest, as it will reflect the "missions" of the departments they head.

(4) Accommodation Model: Suggested most clearly by the writings of Robert Presthus,[6] the "accommodation" model preserves from the "governmental politics" model the notion of differing goals among decision-makers. However, in this case actors engage in cooperative problem-solving rather than conflictual bargaining. Presthus portrayed the Canadian government elites as having similar backgrounds and fundamental values. Thus, while on particular issues actors have different preferences, they are not adamant about them and consciously seek decisions which will reflect a mix of priorities and interests. Decision-making is consensual in technique, if not in objectives. The prototype actor of the accommodation model is the "total government person" as compared with the "Fisheries man" or the "CIDA man." Our research indicates that this model was the single most frequently applicable of the five models described here in the 1968 to 1979 period of Canadian trade policy.

(5) The Individual Model: There are really two variants of this model. The first emphasizes the values, leadership and personal attributes of key ministers and senior officials. To explain decisions, the analyst employing the model would examine the views, style and personal interests of leaders. The individual model depends for successful explanation on evidence that the leader was able to impose his views (or act sufficiently in isolation) in such a way that examination of his interaction with other actors adds little to the explanation. For this reason, it is perhaps most useful at the Prime Ministerial level, although there probably have been occasions when some ministers and even public servants have enjoyed sufficient prestige and power to make the model a successful tool of analysis. The example of C.D. Howe comes immediately to mind in the trade field.[7] But this model is probably characteristic of relatively few real world decisions, given that, as political scientists have often pointed out, power is an asset which must be "spent" frugally to be preserved. Prime Minister

Trudeau's known predilection to collegial or collective decision-making, moreover, seems likely to reduce the number of instances of this model that we will find in the 1968-79 period. On the other hand, because power is spent strategically by successful leaders, we would expect the model to characterize some very important decisions.

A second variant of the individual model focuses on organizational units rather than on individual leaders. Attention is given to the internal procedures, structures and interactions of a government department to explain outcomes, although allowance for some consultation with other organizations is often made.[8] Allison's version of this model was dubbed the "organizational process" model.[9] In the field of trade decisions, probably the closest empirical examples relate to tariff changes in the annual budget, which were traditionally prepared solely within the Department of Finance. However, its explanatory value for the 1968-79 period is limited by the fact that budget preparation had become more interdepartmental than in the past.[10] Another difficulty with the organizational variant is that it may be incomplete in itself, since the processes described by the other models may occur at the intra-organizational level between sub-units. Any experienced member of an organization can, for example, cite instances of positions taken within the department on the basis of rational calculation, bargaining, or accommodation, the last probably being most frequent, because common membership on the "team" encourages cooperative problem solving.

Our research suggests that approximations of each of these models can be observed in the history of Canadian trade decisions between 1968 and 1979. Some of them, however, occurred more frequently than others, and each tends to have been associated with a particular trade policy "sub-system" (to use John Kirton's useful concept).[11] The sub-systems which we shall examine are the tariff setting process between GATT rounds, the Multilateral Trade Negotiations, special import policy, and the anti-dumping and countervailing duties technical sub-systems. By "testing" these models against the record in several policy sub-systems, we hope to uncover, through comparative analysis, some of the conditions for the operation of the processes described by each model.

NOTES: Chapter One

1. Ingo Walter, "How Trade Policy is Made: A Politico-Economic Decision System," in R.G. Hawkins and I. Walter (eds.), The United States and International Markets (Lexington, Mass.: D.C. Heath, 1972) pp. 17-38.

2. J.H. Warren, "Canada's Role in the GATT Negotiations," Canadian Business Review, Vol. 5, No. 2 (Spring 1978) pp. 36-41.

3. Richard E. Caves, "Economic Models of Political Choice: Canada's Tariff Structure," Canadian Journal of Economics, Vol. 9, No. 2 (May 1976) pp. 279-299.

4. Graham T. Allison, Essence of Decision: Explaining the Cuban Missile Crisis (Boston: Little, Brown and Company, 1971).

5. A.W. Johnson, "Public Policy, Creativity and Bureaucracy," Canadian Public Administration, Vol. 21, No. 1 (Spring 1978) pp. 1-15.

6. Robert Presthus, Elite Accommodation in Canadian Politics (Toronto: Macmillan, 1973).

7. See Robert Bothwell and William Kilborn, C.D. Howe: A Biography (Toronto: McClelland and Stewart, 1979). Bothwell's title for an earlier article is a mark of Howe's central role in the St. Laurent Cabinet: Robert Bothwell, "Minister of Everything," International Journal, Vol. 31, No. 4 (Autumn 1976) pp. 693-702.

8. A.W. Johnson, op. cit.

9. Graham T. Allison, op. cit.

10. R.M. Burns, "The Operation of Fiscal and Economic Policy," in G.B. Doern and V.S. Wilson (eds.), Issues in Canadian Public Policy (Toronto: McGraw-Hill Ryerson, 1974) pp. 286-309.

11. John Kirton, The Conduct and Coordination of Canadian Government Decision-Making Towards the United States, unpublished doctoral dissertation, Johns Hopkins University, 1977.

Chapter Two

THE ENVIRONMENT OF TRADE DECISION-MAKERS

We now take leave of decision models until they reappear in Chapter Six. The intervening chapters focus on environmental factors and the major private and governmental actors in the trade policy process.

The trade decision-making process is influenced by a broad environment of political and economic realities and a background of behavioural traditions and past policy choices. This chapter will introduce these environmental factors: the historical and contemporary goals of Canadian trade policy, constraints facing decision-makers, Canada's general trade policy since World War II, and the domestic and international legislation against which specific decisions were made in the 1968-79 period.*

(1) Goals

Governments have manipulated tariffs and other import barriers for a multitude of conflicting purposes, economic and political. A listing of such goals would include employment stimulation, control of inflation, management of the balance of payments, regional income distribution, resource allocation between sectors of the economy, as well as political objectives ranging from electoral expediency to interest group brokerage and national diplomacy.

A rapid glance at the history of Canadian commercial policy reveals that all administrations have pursued simultaneously three broad economic objectives.

* This chapter assembles in one place otherwise scattered information on these subjects. Intended for unitiated readers, it may be safely passed over by those already familiar with the history and law of Canadian trade policy.

First, maximum exploitation of situations of current Canadian comparative advantage has been sought through negotiations to improve access to foreign markets for export industries. A recent variant of this goal was the so-called "sector approach" to the Multilateral Trade Negotiations. This aimed at reducing foreign tariff and other barriers to the upgrading of certain key Canadian resources prior to export. A second economic goal has been encouragement of new areas of potential advantage. Canada's first (and arguably, only) systematic approach to this end was the 1879 National Policy which insulated infant industries such as textiles and iron and steel with a high protective tariff. In the past half century, the protection instrument has been used more sparingly in pursuit of new areas of advantage. This is explained by GATT constraints on raising tariffs, the availability of alternative instruments in the form of fiscal and industrial incentives, and a growing consensus that the Canadian problem is less to establish new industries than to rationalize existing ones. This rationalization is more and more believed to require liberalization rather than protection. However, Canada has yet to abandon in a decisive way the third continuous trade policy goal, that of protecting fields of declining advantage by tariffs and other measures.

It is clear that trade-offs must be made between the pursuit of these economic objectives of trade policy. It is a reality of international negotiation that access to foreign markets cannot be achieved without some opening of domestic markets, and whatever choice is made will impinge on other economic and political goals. Before World War II, most governments emphasized protection of infant and declining industries, particularly since tariffs were one of a few instruments of political patronage. After the war, general policy evolved in the direction of maintaining existing levels of protection while at the same time calling for negotiated multilateral reductions in the GATT framework. The balance struck between economic goals has also varied with prevailing macro-economic and political circumstances and, particularly in the first half-century of Confederation, according to party label - the Liberals emphasizing the interests of export industries and the Conservatives the protection of central Canada's manufacturing industry. But neither party has ignored completely either consideration; differences between the parties were more salient in hustings rhetoric than in their records in office.

Beginning in the 1960s, there have been experiments with a new economic goal: the conscious use of commercial policy to effect rationalization of the problem-ridden Canadian manufacturing industry. In some cases, this latest twist to the goal of developing new areas of comparative advantage, inspired by a series of studies by academic and government economists in the 1950s and 1960s[1], has emphasized liberalization rather than protection as a way of encouraging greater specialization and economies of scale. Although most of the academic advocates of this approach are unhappy with the extent to which governments have implemented the new objective, a few significant steps (e.g., the Auto Pact) have been taken in this direction. The use of commercial policy in the service of rationalization has, in other cases, been pursued through "sector strategies" or packages of industrial, fiscal, research and commercial policy measures that included what was announced as "temporary" protection. The textile, footwear and television strategies of the Department of Industry, Trade and Commerce in the 1970s would seem of this type. "Temporary" protection was justified as necessary to prevent the collapse of industries during the period of rationalization. It of course remains to be seen just how temporary this protection will prove, but is is likely that governments have by now accepted at least the principle advanced by the academics that commercial policy should be used to encourage rationalization of the manufacturing sector, whether this entails liberalization and/or temporary protection. However, they have faced formidable resistance from interest groups, as well as the assumed force of public opinion (regarding trading ties with the United States for example). These constraints have prevented the decisive implementation of a new industrial strategy to replace MacDonald's now somewhat discredited National Policy of 1879.

Government goals of a non-economic character have been another continuous theme in the history of Canadian trade policy. The introduction of British preferences in 1897 had a major effect on overall commercial relations, particularly up to the decision in 1947 to sever the fixed relationship between preferential and other rates of duty. Since 1947 the effect of Commonwealth preferences on Canadian trade patterns has declined in relation to their political symbolism. Since Britain entered the European Community, even the symbolism has waned. In recent years, preferences for developing countries have been a major political thrust of commercial policy. Foreign policy goals, particularly the post-war desire to create a more stable international environment and greater economic linkages between Western democracies, also played a major role in Canada's strong support for the GATT in the 1940s.

(2) Constraints on Decision-Makers

The trade decisions of all Canadian governments have been influenced by constraints on their freedom of action, both domestic and international. Before the Depression, a "link between the interests of the Treasury and the protected industries"[2] was created by the importance of customs duties as a source of public revenues, and this imposed limits on the degree of liberalization acceptable to governments. In all periods, interest group and provincial pressures have been a factor in commercial policy. As Chapter Three will be devoted entirely to the role of private associations and the provinces, at this point we simply take note of their importance.

Historically, the most important domestic constraint on trade policy-makers has been the anxiety of many voters and elites about the consequences for Canadian nationhood of certain kinds of trade agreements with the United States. This fear was the rock upon which foundered the Reciprocity Treaty of 1911 and the behind-the-scenes consideration given in 1947-48 to a Canada-U.S. Free Trade Area. It was also the reason for rejection of the "second option" of the Trudeau government's review of Canada-United States relations, an option pointing to greater continental economic integration.

Canadian trade decisions are equally influenced by constraints associated with the international economic, legal and political environment. In the age of interdependence, countries like Canada, with a small and open economy in which about a quarter of GNP is exported and imported, are particularly subject to exogenous influences. Canada's dependence on external markets makes it more vulnerable than most countries to foreign trade obstacles. This has accounted for generally strong Canadian support for multilateral trade liberalization in the post-war period. The threat of retaliation against Canadian import barriers is probably the single most important constraint upon Canadian protectionism, particularly in relation to the United States, where over two-thirds of our exports are concentrated.

Certain characteristics of the Canadian market work to further limit the freedom of decision-makers. Proximity to the larger U.S. market means that overseas exporters may on occasion flood the Canadian market if trade policy differences between Canadian and American governments given them an incentive to do so. It was in large measure this fear that discouraged Canada from proclaiming its 1972 legislation on preferences for developing countries much in advance of the United States.[3] The high level of direct foreign, especially American, investment in the Canadian market imposes limitations on the ability of Canadian politicians to pursue their cherished trade diversification schemes such as the Trudeau government's Third Option, and the Diefenbaker government's pledge to divert 15 percent of Canadian imports to Britain, since so much Canada-U.S. trade is in the form of transfers between parent and subsidiary companies.

The views and actions of other governments affect Canadian trade decisions in four major ways. Many changes in Canadian tariffs and other barriers are initiated by foreign representations in bilateral and multilateral contexts. Second, actions by foreign governments have implications for Canadian trade performance and, if harmful, lead to decisions to ease the blow. An example is provided by the August 1971 American import surcharge which led to serious re-examination of Canada's "special relationship" with the United States. Third, the posture which other countries adopt toward Canada and its trade problems is significant. For instance, the sentiment prevalent in Europe and the United States that Canada got something of a "free ride" - fewer concessions made than gained - during the Kennedy Round probably influenced the concessions Canada was obliged to make in the Tokyo Round negotiations.[4] Lastly, the framework of rights and obligations built into the General Agreement on Tariffs and Trade legitimizes foreign retaliation against Canadian actions harmful to their interests.

A final set of constraints may be found in the international diplomatic environment. During the 1950s the Cold War atmosphere briefly affected Canadian grain exports to Communist countries, but on the whole Canada's export dependence has discouraged the use of trade policy as an ideological weapon, as evidenced by the refusal to go along with the American boycott of trade with Cuba. Canada has been influenced, however, by less extreme diplomatic influences on trade policy, such as the emergence of developing countries as a force in world politics, and it has participated in UNCTAD commodity negotiations and the Generalized System of Preferences (GSP).

(3) Canada's General Post-War Trade Policy

The most durable tradition of behaviour in Canada's and most other countries' commercial policies has been the principle of reciprocity, with its bias against unilateral liberalization of import barriers. This seldom contested orthodoxy is justifiable more as a negotiating strategy than it is in terms of economic welfare and efficiency. As Ernest Preeg has observed:

> Although advocates of liberal trade are quick to point out the potential benefits of unilateral tariff reduction, this en-

> lightened view has seldom achieved a national consensus. Protectionist resistance to tariff reduction on particular products is difficult to offset by widely dispersed gains in efficiency or consumer welfare.... The bargaining approach, on the other hand, through which gains for export industries are matched against increased imports, has greatly strengthened the appeal of trade liberalization.[5]

Because a country's bargaining power in bilateral and multilateral negotiations is based on the concessions that it is able to make, governments have been most reluctant to "give away" this leverage in advance through unilateral liberalizations. The major Canadian exceptions to the reciprocity tradition were the granting of British preferences in 1897 (later met by British concessions) and the ten year program of preferences for developing countries announced in 1974. There have of course been many examples of non-reciprocal lowering of duties on a temporary basis, but unilateral reductions of statutory or permanent rates have proven quite rare.

The new trading system established after World War II did not alter the policy of reciprocity. But whereas the pre-war period was characterized by bilateral negotiations with the United States and other countries, of which the 1935 granting of intermediate tariff status to American goods and the 1938 agreement on broad reciprocal reductions were important examples, after the war reciprocity became a largely multilateral exercise pursued through GATT negotiating rounds. In historical perspective, the multilateral approach to trade liberalization has proven a grand success. It led to a declining use of tariffs as instruments of protection and an unprecedented period of trade expansion. The Canadian tariff structure has been greatly lowered through seven GATT rounds, although Canada continues to employ this instrument rather more than most industrialized countries.

The accomplishments of the GATT system have no doubt contributed to a much more explicit Canadian commitment to trade liberalization than was generally the case before the war. There have certainly been many bumps along the road to liberalization as travelled by Canada since the war, but by historical standards the commitment seems genuine.

Reciprocal liberalization through multilateralism has thus stood as the basic precept of Canadian trade policy since 1945, and this applied as much to the Trudeau government as to other post-war administrations. It has been a flexible policy however, and several periods of rising protectionism, some of them during the 1968-79 years, have periodically contradicted the general trend.

(4) Specific Canadian Trade Policies: The Domestic and International Legal and Institutional Structure

(a) Domestic Legislation

The British North America Act of 1867 gives jurisdiction over "the regulation of trade and commerce" to the federal Parliament (Section 91.2). However, the fact that revenue legislation can only be introduced by the Governor in Council, as well the the tradition of party discipline, ensures that the federal Cabinet enjoys full control over the tariff and trade policy in general. A number of statutes define the instruments of commercial policy and distribute them between ministers and their departments. The most important are customs duties, non-tariff measures, administrative procedures, review mechanisms and the negotiation of international agreements.

The exercise of the most traditional instrument of trade policy, the tariff, is defined in the Customs Tariff (R.S.C. 1970, C-41). This enormous piece of legislation sets out the various rates of import duty for thousands of products. It further enumerates the countries benefiting from Canada's four tariff schedules: the British Preferential Tariff (traditionally the lowest rates, since 1974 in numerous cases the same as the General Preferential Tariff); the General Preferential Tariff (for developing countries on some

1500 products); the Most Favoured Nation rate (for almost all other countries); and the General Tariff, the highest rate for a few countries not entitled to any of the other schedules. The authority to extend or withdraw any of these rates is invested in the Governor in Council, although in practice this means the Minister of Finance, since tariffs are a form of taxation.

Canada's tariff system contains distinctions beyond the BP, MFN, GPT and General categories which add to the number of rates of duty. Many agricultural products contain seasonal rates that provide protection to Canadian producers from American competition during the domestic growing season. The distinction "of a class or kind not made in Canada" allows lower or free rates for many products. American exporters have frequently complained about class or kind provisions which allow duties on industrial goods to vary between 0-15% according to current Canadian availability of the goods. Separate trade agreements with Australia, New Zealand and South Africa provide for rates on many food items that differ from British Preferences: canned beef for instance has a 15% BP rate, but enters free under the Australia-Canada agreement. Provision is made in Schedule B of the Customs Tariff for drawback or remission of duties on imports used as inputs into Canadian manufacturing. Raw sugar, for example, can receive a drawback of 99% if refined in Canada for use in wine making.

The Customs Tariff also authorizes the Governor in Council to levy a countervailing duty on goods deemed by Cabinet to have been subsidized by foreign governments (Section 7), and to impose surtaxes on products imported from a country which discriminates in its tariff schedules against Canada (Section 8.1) or against products deemed to be "causing or threatening ... serious injury to Canadian producers" (Section 8.2: based on the Article 19 "safeguard clause" of the GATT).

Finally, the Government's authority to take part in multilateral and bilateral trade negotiations is contained in Section 11: the Governor in Council may reduce duties "by way of compensation for concessions granted" to Canada by other countries. The schedules of the Customs Tariff are subject to frequent change. Cabinet may temporarily reduce duties, by order in council for products used as inputs into Canadian manufacturing, and by legislation (usually the budget) for any other duties.

Turning to non-tariff instruments, provision for quantitative restrictions on imports and exports is made in the Export and Import Permits Act (R.S.C. 1970, E-17), originally passed in 1947. The Governor in Council, upon recommendation of the Minister of Industry, Trade and Commerce, is empowered to place goods on an Export Control List for reasons of national security, availability of supply or international agreement (such as commodity agreements), and on an Import Control List for reasons of scarcity of supply, agricultural price stabilization in Canada, international agreement (e.g., commodity agreements like tin and coffee, where Canada is a consumer member, to prevent or limit imports from non-member countries), or to prevent "serious injury or threat of injury" pursuant to the deliberations of the Textile and Clothing Board or the Anti-Dumping Tribunal (Section 5.2). An Area Control List was also established by the Act to control or prohibit the import of goods from certain countries, such as Rhodesia in accordance with United Nations sanctions. Quantitative controls are administered through the issuance of permits to importers and exporters by the Ministry of Industry, Trade and Commerce. Most cases of export control have involved strategic goods, while the Import Control List has been used primarily to limit imports of dairy and meat products from the United States and New Zealand, and "low cost" manufactures (e.g., textiles, footwear).

Quantitative restrictions may also be based on the general trade and commerce power of the federal executive. Voluntary export restraint agreements (VERs), sometimes called "orderly marketing arrangements," may be negotiated with foreign exporting countries, which then "voluntarily" limit export of given products to Canada to agreed levels. Numerous examples exist in the area of textiles.

Another group of non-tariff instruments affecting trade flows is the myriad of statutes, regulations and administrative decisions by a host of federal and provincial ministries, which obstensibly have some other purpose than limiting imports, but which may consciously or unconsciously have trade implications. Some of these non-tariff measures, such as labelling requirements and health and safety regulations under the Food and Drugs Act, the Hazardous Products Act and other statutes, involve border action and are administered by the Department of National Revenue, Customs and Excise Branch, at Canadian ports of entry. Other actual or potential non-tariff barriers (NTBs) include federal and provincial government procurement regulations, regional development grants under the Department of Regional Economic Expansion, and programs of adjustment assistance under the Department of Industry, Trade and Commerce.[6]

Many of the administrative procedures of trade policy are outlined in the Customs Act (R.S.C, 1970, C-40) which authorizes the Governor in Council to make regulations prescribing the terms and conditions under which goods may be imported into Canada. Sections 36 to 44 contain the value-for-duty provisions describing the value base of goods on which rates of duty are to be applied (defined in Section 39 as the "fair market value at the time when and the place from which the goods were shipped to Canada"). The administration of the Customs Act and the entire Customs service is under the responsibility of the Minister of National Revenue (or Revenue Canada). The Customs Act, however, is more than just administration. It can be used to effect what are generally known as administrative barriers to trade. Although it seems clear that the extent to which Canada has engaged in such practices has declined since World War II, many countries have complained of protectionist intent to some Canadian customs practices such as its non-adherence to the almost universal "Brussels" system of product classification and the "fair market" valuation procedures.[7]

In accordance with the post-war GATT aim of avoiding arbitrary restrictions on trade, three major review mechanisms have been set up by the Canadian government to determine the facts of certain disputed questions and to make recommendations to ministers. These bodies, to be discussed more fully in Chapter Four, are the Textile and Clothing Board, the Tariff Board and the Anti-Dumping Tribunal.

A final instrument of trade policy is the ability to enter into international trade agreements, both bilateral and multilateral. This derives from the general treaty-making power of the Governor in Council, from Section 11 of the Customs Tariff, and from Section 4(c) of the Department of Industry, Trade and Commerce Act of 1968 (R.S.C. 1970, I-11), which requires the Minister of IT&C to "improve the access of Canadian produce ... into external markets through trade negotiations...."

(b) International Legal and Institutional Framework

Canada is party to a number of international treaties, agreements and institutions which carry with them rights and obligations, and thus involve both opportunities and constraints for policy-makers in Ottawa. Some of these accords are multilateral (the GATT, several international commodity agreements, the Arrangement Regarding International Trade in Textiles); others are bilateral (trade agreements with individual countries or groups of countries, the Automotive Agreement and the Defence Production Sharing Agreement with the United States). Although they are not based on legal requirements, participation in certain institutions such as UNCTAD and OECD does have some bearing on the making of Canadian trade decisions.

By far the most important trade accord to which Canada adheres is the General Agreement on Tariffs and Trade (GATT), signed in 1947 by 23 original contracting parties, including Canada. Its aims, as set out in the Preamble, are to encourage the "substantial reduction of tariffs and other barriers to trade and the elimination of discriminatory treatment in international commerce." The GATT is composed of four parts:

Part One outlines the fundamental responsibility of GATT members - the Most Favoured Nation or MFN principle, which obligates governments to extend any reduction of duties accorded one member to all members (Article 1). Permitted exceptions to the MFN rule include preference systems already in existence when the GATT was signed, as well as free trade areas and customs unions, provided they do not increase barriers vis-à-vis outside countries.

Part Two of the GATT amounts to a code of fair trading behaviour in which the rights and obligations of states are set out. Obligations on governments include the prohibition of internal taxes that discriminate against imports and customs valuation based on domestic prices, as well as the provisions that restrictions imposed on imports should be non-discriminatory, and that independent review and judicial mechanisms should be established. The rights of states are outlined in the articles defining conditions under which restrictions on imports and exports are permitted. Countervailing and anti-dumping duties are recognized under Article 6. Temporary restrictions are allowed to protect the balance of payments (Article 12) and, most important, to restrict imports on an emergency basis when these imports "cause or threaten serious injury to domestic producers" (Article 19 - the "safeguard" clause). Although it is not found in Part Two, Article 28 establishes the right of compensation to members when another party increases bound rates of duty.

Part Three of the Agreement essentially outlines the operating rules of the GATT. Article 25 allows a waiver of any obligation under the agreement if the state in question can get approval of two-thirds of the membership. Article 28 bis 1 provides for periodic negotiating conferences or "rounds" to take place for the purpose of further liberalization of trade. Article 28 bis 2 recognizes the interests of low-tariff countries by considering the binding of low or free rates to be a concession equivalent to the reduction of higher duties.

By Part Four of the GATT, ratified in 1965 in response to the demands of the developing countries, developed countries undertake to reduce barriers to Third World exports "to the fullest extent possible," without expectation of reciprocity.

These provisions of the GATT constitute a major constraint upon Canadian and other trade decision-makers. Although the vagueness of many articles and the several escape clauses allow great flexibility on the part of national governments, the agreement does force many specific obligations on member states. In 1962, for example, Canada was requested to revoke the emergency surtax applied to potatoes by a panel set up under Article 23 and was given a slap on the wrist for its import surcharge. Article 28, the principle of compensation for tariff increases to trade partners, causes decision-makers to think twice about raising duties. Another constraint on policy-makers is psychological: a tradition of reluctance to break the "spirit" of the agreement has grown over the years. And, more concretely, the "spirit of the GATT" provides ammunition to opponents of contravention within governments.

Canada's participation in multilateral accords also includes several international commodity agreements. The most important of these are the International Tin Agreement, which operates a buffer stock to stabilize tin prices and to which Canada contributes as a consuming country, the International Lead and Zinc Study Group, which draws up supply and demand forecasts in an attempt to moderate price fluctuations, the International Cocoa Agreement, the International Coffee Agreement, and the International Sugar Agreement. Under the Food Aid Convention of the current International Wheat Agreement, Canada is committed to supplying grains to developing countries. The present wheat agreement does not provide for market intervention to influence prices.

Another multilateral trade agreement to which Canada adheres is the Arrangement Regarding International Trade in Textiles or Multifibre Agreement (MFA), an attempt to inject some order into the ways in which industrialized countries deal with surges of "low cost" imports from developing and East European countries. The Agreement is unique in that it allows discrimination (unlike the GATT) in the application of special restrictive measures against textile imports.

In addition, there is a host of bilateral trade agreements between Canada and other individual countries. Most of these involve exchange of most favoured nation treatment or, in the case of most Commonwealth countries, exchange or accordance of British preferential treatment. Two sectoral agreements with the United States, the Automotive Agreement and the Defence Production Sharing Agreement, provide for free trade in these products. Although they involve no commercial concessions, the Canada-EEC Framework Agreement for Commercial and Economic Cooperation (June 1976) and the Canada-Japan Agreement (October 1976) provide for closer consultation and potential cooperation on trade matters.

Canada is also a participant in the activities of the United Nations Conference on Trade and Development (UNCTAD). In contrast to the GATT, UNCTAD is not a set of contractual obligations on members, but rather an ongoing forum for the discussion of trade issues as they apply to developing countries. The practical importance of UNCTAD to Canadian trade policy is illustrated by the fact that some obligations have in the past had their origins in UNCTAD Conferences (e.g., the Generalized System of Preferences, the International Cocoa Agreements of 1973 and 1976). UNCTAD is now the principal negotiating forum for international commodity agreements. The activist ideology of the UNCTAD Secretariat is also relevant. Its constant scolding of Western Governments has led to some concessions to developing countries by Western countries, including Canada. A series of conferences on individual commodities under the auspices of UNCTAD and an attempt to deal in a comprehensive way with commodities via the so-called "integrated commodities program" are planned for the late 1970s and early 1980s. Canada is also a member of the Organization for Economic Cooperation and Development (OECD), the major forum for coordination of developed countries' approaches to UNCTAD activities.

NOTES: Chapter Two

1. Typical of these studies was: H. Edward English, Industrial Structure in Canada's International Competitive Position (Montreal: Private Planning Association of Canada, 1964).

2. John H. Young, Canadian Commercial Policy, a study for the Royal Commission on Canada's Economic Prospects (Ottawa: Queen's Printer, 1957) p. 43.

3. An Act to Amend the Customs Tariff (S.C. 1973-74, 10).

4. This argument is made, for instance, in Harold B. Malmgren, "The Evolving Trading System," in H. Edward English (ed.), Canada-United States Relations, Proceedings of the Academy of Political Science, Vol. 32, No. 2 (Montpelier: Capital City Press, 1976) pp. 125-136.

5. Ernest H. Preeg, Traders and Diplomats (Washington: Brookings Institution, 1971) pp. 23-24.

6. The most complete analysis of Canadian non-tariff barriers can be found in Klaus Stegemann, Canadian Non-Tariff Barriers to Trade (Montreal: Private Planning Association of Canada, 1973).

7. The "fair market" valuation system applies duties to the home price of a product rather than to its export price. Sometimes the export price is lower than the home price, for such "legitimate" reasons as small domestic markets in the supplying country and differences between distribution systems in Canada and the country of origin. On other occasions, the lower export price may reflect less legitimate practices like dumping end-of-line or end-of-season products (e.g., clothing) on the Canadian market. When Canada Customs officers apply rates of duty to the higher home price, Canadian producers get added protection.

Chapter Three

PROFILES OF TRADE POLICY ACTORS: INITIATING ACTORS

An indication of the importance of trade in the Canadian economy is provided by the very large number of actors or organizations involved in the policy process. Interest groups, provincial governments, Members of Parliament, government departments and Cabinet ministers are all participants, although some with greater regularity and intensity than others. It is not difficult to find examples of decisions in which actors from any one of these categories played crucial roles.

We shall divide the presentation of these principals and their characterisitcs into three chapters on "initiating actors," "specialized review actors" and "departmental actors." This categorization is intended to correspond to a simple model of the structure of government decision-making sequences in Canada.[1] With some simplification, any decision-making sequence in Canada may be divided into four basic phases: an initiation phase in which an issue arises (to take an obvious example, an interest group demands protection); a review phase in which one or more actors analyse the facts of a problem, interpret it and present options to the appropriate authority for decision; a bureaucratic resolution phase in which officials from one or more departments develop recommendations to Cabinet; and a political resolution phase in which final choices between options are made in Cabinet. On trade issues, the major players in the intitiation stage are interest groups, provincial governments, and Members of Parliament. The review phase may be undertaken by both departmental officials and what we term the "specialized review actors," the Tariff Board, Textile and Clothing Board and Anti-Dumping Tribunal, which we have singled out for treatment in a separate chapter. An introduction to the participants in both the bureacratic and political resolution phases is provided in Chapter Five, on government departments and central agencies. We recognize that some overlap exists between the actors involved in these stages. For example, departmental bureaucrats are often initiators and reviewers as well as producers of recommendations to ministers. Our purpose is simply a convenient, albeit imperfect, categorization of actors based on their most characteristic functions in the trade policy process.

(1) Interest Groups

Trade policy, particularly the tariff, was probably the original focus of interest group demands on Canadian governments. S.D. Clark's classic study of the Canadian Manufacturers' Association argued that MacDonald's National Policy tariff of 1879 was virtually set by the Government asking the Association what it wanted, and giving it to them.[2] It is much more difficult today for any one interest to have its way, even those with great resources and commanding large blocs of votes. An unprecedented process of economic and social differentiation has unfolded during the 20th century, and this has been associated with a proliferation of interest groups. One of the most important checks on the unbridled influence of any one interest is the existence of other interests. This holds not only between sectors (e.g., consumer vs. manufacturer) but also within sectors (e.g., one manufacturer wanting added protection is opposed by another who uses the item as an input). In addition, governments during this century have armed themselves with a multitude of policy instruments, including tax policy, regional expansion policies, industrial adjustment incentives, etc., so that instruments affecting trade flows are today only one of several tools available to satisfy interest group demands. Finally, the post-World War II institutionalization of certain international norms of acceptable trade behaviour, by the GATT especially, has constrained the ability of governments to satisfy protectionist group demands.

While these factors have made it more difficult today for any one interest group to have its trade demands met by governments, it is true nevertheless that these groups and their behaviour are of enormous importance to the making of trade policy, and also that some are unquestionably more powerful than others. Probably the majority of individual trade decisions have their origins in demands on governments from some domestic pressure group. Moreover, even when decision sequences are initiated by other actors, such as government officials or foreign governments, interest groups play the important role of providing information to governments as to what needs to be negotiated with other governments, and on the technical and political feasibility of options. For politicians especially, this information flow function is crucial. In the mass society interest groups are an essential link between government and the electorate.

What is less desirable (though difficult to avoid) is that some interests intervene in the policy process with far greater resources than others. In Canada's trade policy, the resources which most differentiate interest groups from one another are the capacity to command electoral support and the capacity to furnish sophisticated technical information. The distribution of these resources among Canadian interest groups appears to have a protectionist bias on the making of Canadian trade policy. It is the groups with an interest in maintaining levels of protection (but not necessarily increasing them) that command the most political support, although some groups with liberal interests, such as the resources industries and Western agriculture, are far from powerless. The concentration of internationally uncompetitive manufacturing industries and marginal agriculture in densely populated Ontario and Quebec, and the greater intensity of opposition by producer groups to adverse government trade decisions as compared with consumer groups, is central to this asymmetry of influence. Because these industrial and agricultural groups are powerful, they are generally in a position not only to intervene on issues with more frequency but also to provide more sophisticated information to government decision-makers.

(a) The Trade Interests and Views of Major Canadian Industries or Sectors

The occasion of the recently completed Multilateral Trade Negotiations or MTN ("Tokyo Round") provides a convenient opportunity for the study of pressure group demands on Canadian trade policy makers. Many of these groups' briefs to the Canadian Trade and Tariffs Committee (CTTC) have been made public, and collectively they provide a snapshot of the trade problems and views of major industries in the late-1970s.

(i) Agriculture

Canadian agriculture is far from homogeneous in its trade interest. The cereal and feed grains sector is already highly export-oriented and its major interest lies in liberalizing foreign barriers to grain exports, particularly in the EEC and Japan, and in price and reserve stabilization. At the other end of the spectrum, the dairy and the fruit and vegetable sectors have proven vulnerable to foreign competition and have been sheltered by quotas and seasonal tariffs for many years. In a more intermediate position lies the livestock sector. The beef industry has historically been quite healthy in Canada, but periodic import and export crises have occurred, such as those in 1972 when Canada temporarily limited imports from the U.S. and in 1977-78 when American cattlemen blockaded the border and demanded protection from their government. The pork, poultry and lamb sectors have been more consistently subject to import pressure: turkeys, eggs and veal are presently on the Import Control List.

The agricultural sector, moreover, is extremely sensitive politically due to the regional concentration of production (e.g., grains on the Prairies, dairy products in Quebec and Ontario, fruits and vegetables in Ontario and British Columbia), the recent organized consumer interest in food problems,[3] and those aspects of agricultural policy which aim at stabilization of farm incomes. Moreover, Canadian markets for many products have been managed by governments and this has necessitated control of imports to prevent them from undoing current government price policies.

The major interest groups in agriculture, generally quite effective lobbies, are the Canadian Federation of Agriculture, an umbrella group, and the sectoral associations like the Canadian Cattlemen's Association, the Canadian Horticultural Council, the Canadian Pork Council and the Dairy Farmers of Canada. The Federation, in its representations to the CTTC, found it difficult to reconcile the competing trade interests of its various regional and sectoral components. The major demands of its briefs were the protection of the weak sectors from foreign (primarily American) competition, the opening up of markets in the major industrial economies, and the lowering of tariffs on many inputs used by farmers. The first two of these desires are of course difficult to achieve in international trade negotiations; negotiating access to the EEC for grain exports, for instance, will not be eased by maintaining or expanding protection of the dairy sector.

(ii) The Primary Resources Sector

The forest products, fisheries and minerals sectors export into highly competitive international markets, and are supportive of trade liberalization, particularly improved access to American, European and Japanese markets. The major exceptions are the plywood and fine papers industries, and canned tuna and shellfish, which are protected at home and abroad by tariffs.

The major primary industry associations are the Council of Forest Industries of British Columbia, the Canadian Pulp and Paper Association, the Fisheries Council of Canada and the Mining Association of Canada. These groups also play a major role in the Canadian Export Association. Their importance to Canada's trade performance is illustrated by the fact that the government's original negotiating strategy for the Multilateral Trade Negotiations, the so-called "sector approach," was largely constructed around the interest of these sectors in reducing foreign escalating tariffs which hinder upgrading of Canadian resources prior to export.

(iii) Manufacturing

The problems of the Canadian manufacturing sector, especially its sorry export and import competition performance, are well documented. In addition to short-term problems, such as comparative wage rates and the level of the dollar, certain trade-related problems were identified by the Economic Council in 1975:

> One of the basic causes of our poor productivity performance is the type and organization of manufacturing fostered by the commercial policies adopted by Canada and other countries over the years.... In Canada, in contrast with primary or resources industries, most of the manufacturing sector has been unable to achieve economies through access to larger external markets because both Canada and its major trading partners have provided much more protection for highly manufacturered products. At the same time, Canadian protection has allowed manufacturing firms operating in Canada to increase their prices behind the tariff and so maintain otherwise uneconomic production runs.[4]

To be sure, the competitive position of individual industries varies greatly. At one extreme lie a number of severe problem industries, such as textiles, clothing, footwear, furniture and some standard electrical items, where the above problems are compounded by sharp wage differentials between Canadian firms and enterprises in developing countries. These sectors, largely located in populous Quebec and Ontario, have been in the vanguard of protectionist tendencies in Canada. At the other extreme lies a group of competitive manufacturing industries, such as telecommunications, automobiles, pulp and paper, and agricultural machinery, which have held this enviable position through a combination of resource-intensive production and low international (or North American) trade barriers. In the middle lie numerous industries which are potentially competitive, if industrial rationalization (in terms of specialization of firms and product lines) and more dynamic research and development practices can be developed.[5]

Many Canadian firms, it should be noted, have mixed trade interests, because they are often simultaneously import-competing on their end product but large importers of inputs into their production processes. Examples are the clothing and food processing industries which desire protection against imports of their end products but low barriers on textile fabrics and fruits and vegetables. Like most countries, Canada's tariff structure contains an element of escalation as the degree of processing of a product increases.[6]

The Canadian Manufacturers' Association (CMA), which represents about 75 percent of firms in this sector, has traditionally opposed tariff reductions, partly because it is dominated by its less competitive members. In its two submissions to the Canadian Trade and Tariffs Committee, the CMA opposed the use of a general tariff-cutting formula and argued that "the few remaining restraints on imports into Canada are justified"; however, it did not propose increased tariffs.[7] It also called for the termination of preferential rates on British imports, opposed adoption by Canada of customs valuation based on export prices and argued against any Canadian concessions on government procurement until trading partners "indicate their willingness to adopt a (similarly) liberal policy."[8]

(iv) Exporters, Importers and Retailers

These groups constitute the core of the "trade-biased" sector, to use Walter's terminology. The Canadian Export Association (CEA) is the strongest of the three groups and advocates trade liberalization, but the intensity of its commitment is modified somewhat by the diverse composition of its membership. In addition to the primary industries with a clear export orientation, the CEA's 400-odd member companies include many secondary manufacturing firms with only a marginal export orientation or which have certain product lines which depend on the protective tariff for survival. As a result, the CEA's MTN recommendations[9] emphasized the reduction of foreign barriers to Canadian exports (especially some United States policies such as the "buy American" act, countervailing and anti- dumping legislation) but made concessions to its less competitive membership by de-emphasizing tariff reductions. The CEA is considered a highly competent and effective lobby in Ottawa circles. This is not surprising, given the importance of these generally dynamic companies to the Canadian economy.

The Canadian Importers Association (CIA), probably the group with the clearest interest in free or freer trade, has been in more of a conflict relationship with government, particularly over quantitative restrictions on textile imports and the administration of the anti-dumping and valuation regulations by Revenue Canada. The CIA represents only a minority of all Canadian importers. Like many of the relatively weak interests it has tended to be rather shrill in its argumentation and publicly critical of trade decisions. Its credibility in federal government circles has probably been limited by these tactics and by the knowledge that it represents only a few thousand people. Canadian retailers, represented primarily by the Retail Council of Canada and the Canadian Chamber of Commerce, are also among the most free trade-oriented groups in the country.

(v) Consumer Organizations

The Consumers' Association of Canada (CAC) has intervened periodically to defend the interest of consumers, especially those with low incomes, in keeping import barriers to a minimum. These interventions have concentrated on high profile import matters such as quantitative controls on textiles and agricultural products (e.g., eggs, turkeys) and tariffs on sugar and fruits and vegetables. The Association's impact has been limited by inadequate financial and research resources as well as by periodic internal divisions. On some of the issues involving imports from developing "cheap labour" countries, for example, the Quebec wing of the association, and many individuals in the 100,000 strong movement in other parts of the country, have proven sympathetic to local labour and business appeals for protection, and undermined the position of the national organization. While strongly committed to trade liberalization, CAC has tended to relegate the international aspects of consumer protection to second place behind the domestic questions of competition policy and product standards. Perhaps its greatest influence has been in promoting health and safety, product standards, and packaging and labelling regulations which sometimes constitute non-tariff barriers to imports. One of the most important roles of consumer organizations is the simple fact that their interests can be invoked by policy-makers wishing to resist protectionist demands.

(vi) Labour

The influence of organized labour on Canadian trade policy is exerted primarily at the local and industry union level, via pressure on Members of Parliament, and in unison with management in most cases of import competition. The traditional labour-management alliance on trade issues is less apparent, however, in multinational enterprises, which have been known to shift production abroad in response to import pressure rather than to lobby for higher protection in Canada.

The labour movement, as a whole, has not proven a very effective influence. The Canadian Labour Congress (CLC) has taken a position in favour of further trade liberalization, but many local or industry sector unions have openly criticized this stance. Its influence is also weakened by the political partisanship of the CLC and its adversary tactics that, for example, led the Congress to withdraw from many government-private sector consultations during the recent period of wage and price controls (1975-78). Perhaps the most useful contribution of organized labour to the trading environment has been in restraining the growth of protectionist sentiment in American labour organizations through their participation in international unions. It is difficult to gauge this influence precisely, but in some cases, for example elections in the United Automobile Workers, it was not inconsequential.

(b) The Behaviour of Interest Groups and Interactions with Government

There is a growing literature on the role of interest groups in Canadian politics.[10] Although we are aware of no specific case studies of interest group lobbying on trade issues (Clark's 1939 study of the Canadian Manufacturers Association perhaps comes closest)[11], our evidence indicates that their behaviour does not markedly differ from lobbying on most other issues. This should not be surprising, given that the groups agitating on trade are pretty much the same organizations that have been studied in other issue areas. The following comments seem appropriate:

(i) Interest groups are an integral part of the political process in Canada, and are generally regarded by bureaucratic and political policy makers as legitimate and positive forces. For politicians these groups are a day to day link with the electorate. For bureaucrats they are a source of both technical and political information on policy options.

(ii) The national associations are generally federations of provincial or sectoral groups, which often hold divergent views, and some reconciliation of differences must be done within the federations themselves if they are to be successful. The most effective lobbying is usually done by the industry-specific associations as opposed to the umbrella groups. Industry specific groups usually have better documented cases and often have one or more Members of Parliament on their side in the Government caucus. Umbrella groups concentrate on exercises, like the MTN, where whole sectors of the economy are on the table. With a few exceptions, such as the Canadian Labour Congress, they avoid political partisanship. The groups active on trade issues vary greatly in resources. The Canadian Manufacturers' Association is equipped to provide numerous trade services to its members (e.g., customs and export procedures) and is able to make interventions quickly. By contrast, a group such as the Consumers' Association has only one officer working on trade issues, with the result that interventions are relatively infrequent and not fully documented.

(iii) Presthus found that the targets of interest group influence attempts are concentrated in the bureaucracy and in Cabinet, with Members of Parliament and Senators a distant third.[12] On trade issues, interest groups generally have a good "feel" for what questions should be brought to the attention of ministers, bureaucrats and Members of Parliament. It is primarily the local, smaller firms and other interests which lobby MP's. Bureaucrats are more likely to be targets for questions which are relatively less sensitive politically (such as the more routine requests for temporary tariff reductions, drawbacks, countervailing and anti-dumping duties). The most sensitive issues (like requests for quantitative controls and changes in statutory duties) tend to be brought directly to the attention of ministers, often through their executive assistants. These are of course only general tendencies and there are many exceptions.

(iv) Interest groups have used numerous methods of communicating their claims to decision-makers. The most important method is probably the formal or informal brief to the responsible minister, the sectoral department minister or the minister with regional responsibilities. Joint briefs between allied groups seem particularly effective, as evidenced by the joint representations of clothing and primary textile manufacturers in the period leading up to the November 1976 clothing quotas (see Chapter Six). Trade policy is one area of public affairs where there has been a great deal of institutionalization of the lobbying process - through such mechanisms as the Textile and Clothing Board, the Tariff Board, the Anti-Dumping Tribunal and, for multilateral negotiations, the Canadian Trade and Tariffs Committee. Many departments, moreover, have advisory boards to communicate industry claims to ministers.

(v) What determines the degree to which the competing interests get their way? It should be understood that the Minister and his senior advisors are in a difficult position on trade policy. They are subject to many vocal and competing domestic demands. Moreover, Canada's official trade policy is one of liberalization and thus any introduction or strengthening of protective actions is subject to GATT legal constraints and retaliation by our trade partners. Policy-makers therefore need some principles upon which to make these difficult decisions. One thing is clear; groups which are perceived as not understanding the policy-maker's delicate position, and as attempting to embarass policy-makers or enter into partisan politics, are at a distinct disadvantage. Many in Ottawa believe for example that the Consumers' Association has frequently committed these errors. Similarly, poorly documented cases usually will be rejected quite easily. But probably the most important criteria for choosing among group claims are the political strength of the groups involved on the one hand and the anticipated international costs of any given action on the other. On the count of political strength, it is probable that groups which represent or can obtain the support of large blocs of voters (such as import-competing manufacturers, compared with the export-oriented resource sector) have an advantage, particularly when the voters are strategically located. Campaign contributions to political parties (or their withdrawal) are always a latent consideration. Intensity of representations is also important. Politicians are clearly more sensitive to potential lost jobs due to import competition than they are to a group like the Consumers' Association which claims to represent all Canadians, but which in the end represents very few; decision-makers know that few people vote in their roles as consumers, at least in most circumstances. Internal divisions within interest groups are also fatal, since, in view of the difficult position of policy-makers, they will take advantage of this opportunity to reject demands.

But while it is probably true that the "protection-biased" sector has greater political or electoral strength than groups supporting trade liberalization in Canada, one should not conclude that this leads to unbridled protectionism. Groups of a liberal bent are by no means without power; nor are their defenders in Cabinet and the bureaucracy. Furthermore, increases in levels of protection are severely constrained by the fear of retaliation by trade partners and by the constraints of international law on tariff increases.

Certain conditions determine the ease with which protection claims are accepted or rejected. Minority governments, or governments facing elections in the near future, often swing in the direction of protection. General economic conditions are important. In periods of rapid growth firms find it easier to adjust to import competition. Inflation is usually a paramount concern; governments then find it much easier to resist protectionist demands. The context of decision-making is also relevant. In multilateral trade negotiations, because all the cards are on the table and governments can play interest groups off against each another, the ability of policy-makers to base their calculations on national as opposed to particularist interests is strengthened. In isolated one product decisions, on the other hand, the situation is likely to be dominated by the protectionist forces, a situation which John Kirton aptly referred to as the "fragmentation effect."[13]

(vi) We have just described the general features of interest group-government relations and the factors which in a given case tend to swing the balance in the direction of either protection or trade liberalization. In the end, policy outcomes seldom amount to complete victory for any one interest. Although there do exist crisis-type situations, such as the one described in our case study on clothing imports, in which one group manages to achieve most of it aims, in the overwhelming number of instances, policy outcomes reflect compromises between the claims of interest associations. For some of the reasons outlined above, in the isolated case the advantages seem to lie somewhat on the side of protectionists, either in the form of increased protection (particularly non-tariff barriers rather than customs duties) or most frequently, maintenance of existing levels of protection. Consequently, liberalization of trade barriers tends to occur mainly through periodic international trade negotiations, where the "fragmentation effect" can be neutralized, and in those happy, but all too few situations of economic expansion.

(2) The Provinces

Provincial participation in trade issues has increased substantially in the past decade. Only three provinces, Nova Scotia, Newfoundland and Ontario, made representations to the Canadian Trade and Tariffs Committee (CTTC) during the Kennedy Round preparations of the mid-sixties. In contrast, not only did all provinces brief the CTTC of their concerns for the Tokyo Round, but extensive and in-depth consultations took place between federal and provincial senior officials and ministers. The explanation for this rising provincial interest in trade issues lies largely in the growing concern in the provinces over industrial development and diversification, which is now clearly seen to be affected by both Canadian and foreign trade barriers.

The Western and Maritime provinces traditionally have supported trade liberalization, while Ontario and Quebec demanded protection for their weak manufacturing sectors. Today, Ontario's interests are more mixed than those of most provinces. The desire for free trade remains particularly strong in the West, where it has long been argued that the costs of tariff protection weigh heavily on Western consumers and discourage diversification of the Western provinces' economies. While Western provincial governments have not been timid about requesting protection for such weak industries as plywood in British Columbia and clothing in Manitoba, in general, Western industries are export-oriented and competitive in world markets. Maritime provinces, while also responsive to the consumer exploitation argument, have been less publicly hostile to Ottawa trade policies, probably because of their dependence on federal transfer payments.

Perhaps the most common provincial behaviour on trade issues is to act as a conduit for the communication of local interest group desires to the federal government. For example, both the Quebec and Manitoba governments, which sit on the Minister of IT&C's Textile and Clothing Advisory Committee, invariably support the industry's demands for increased protection via the Import Control List. Provincial interventions of this sort are encouraged by the fact that there are few political disincentives to taking rather narrow positions supportive of local industries. Generally speaking, the provinces do not have very diverse economies, most being either resource-based or dominated by manufacturing. In addition, there often exists a community of interest on trade issues between management and labour at the local level. Insofar as the difficult trade-offs between economic sectors and socio-economic groups are concerned, the buck is passed to Ottawa.

There is some recent evidence of change. As the desire for industrial diversification grows, the process of reconciling opposing interests is beginning to occur somewhat more at the provincial level than before. The traditional Western free trade bias, for example, is becoming less strident as these provinces request infant industry protection for their nascent manufacturing sector. But these recent changes do not appear to have encouraged provinces to think in terms of conflict resolution at the national level. Provincial governments continued to take rather narrow positions before the Canadian Trade and Tariffs Committee and in the Federal-Provincial Committee of Deputy Ministers of Industry set up to facilitate consultation between Ottawa and the capitals on the MTN. Each province demanded liberalization of foreign barriers for its resource industries and protection for its infant or declining industries. Ontario's submissions, for instance, simultaneously proposed United States tariff cuts on Ontario's industrial and resource exports and the maintenance, virtually intact, of the Canadian tariff structure, supplemented by tinkering with the value-for-duty provisions of the Customs Act to protect fruits and vegetables, and further safeguards in the Auto Pact to maintain production in Ontario.

Also increasing is the direct provincial policy role on trade issues. The provinces have long had a direct involvement in the erection of some of Canada's most conspicuous non-tariff barriers, most notably in procurement for their liquor control boards and public utilities. More recently, provincial treasury and industry departments, especially in Ontario, Quebec and Alberta, have greatly upgraded their trade expertise and participated actively in preparations for the Multilateral Trade Negotiations. Other initiatives have occurred in the area of export promotion, including the establishment of trade offices abroad and direct approaches to foreign governments, of which one of the more visible examples was Premier Regan's 1973 mission to Washington to protest U.S. countervailing duties on Michelin Tire of Nova Scotia.

(3) Parliamentary Actors - Parties, Members of Parliament, Senators

Although the making of trade policy is highly political, the cleavages involved tend to be local and regional interests versus some conception of the "national interest," rather than between political parties. Up to the Depression, the two major parties were distinguishable, at least in rhetoric. The Liberals were identified with Free Trade and United States-Canada reciprocity and the Conservatives with the high tariff wall of Macdonald's National Policy and British preferences. John Weir has argued that even in this period, electoral rhetoric aside, the behaviour of the parties when in power was not all that different.[14] The National Policy was in Macdonald's mind a second-best solution implemented after he failed to win American support for reciprocity at the Washington Conference in 1871, and which he hoped would be a bargaining tool to achieve the desired goal of reciprocity. It was Laurier the Liberal who introduced British preferences in 1897 and the Tory R.B. Bennett who both raised the tariff to its highest level at the outset of the depression and negotiated major cuts with the United States after the 1934 United

States Trade Agreements Act. After the Depression, even much of the rhetoric subsided. Weir observed:

> After the 1930's, the tariff ceased to be a political issue in Canada.... The parties seemed to accept the existing level of protection. As Eric Kierans has put it, "During election campaigns, the Liberals insist that tariffs cannot be raised, and the Conservatives insist that they cannot be lowered."[15]

Today, what seems most to distinguish the parties at any moment is whether or not they are in power. Parties in power have at least attempted some reconciliation of sectoral and regional conflicts in accordance with their view of the "national interest" at that time. Opposition parties tend to cite individual cases and to take the side of local, protectionist forces, from which some political hay can be made. Both parties shy away from such dramatic departures as Canada-United States free trade, no doubt recalling the defeat of the pro-Reciprocity forces in the election of 1911.[16]

The present role of Members of Parliament is consistent with these trends. Both opposition MPs and Government back benchers, with but a few exceptions, tend to intervene on trade matters to defend local interests threatened by import competition. There have been a few individuals in the House who have consistently defended some broader concerns such as the consumer interest and promotion of a consistent industrial strategy, or have expressed internationalist viewpoints, but they are exceptional. The lack of consistent trade ideologies among Canadian political parties is well illustrated by the not uncommon practice of opposition members arguing for higher protection for some industry and then afterwards slamming the government for associated rises in consumer prices, or want of an industrial strategy.

Members do enjoy some influence over trade decisions. This stems from their role as conduits for interest group demands (particularly small local interests rather than the large associations), and in a more limited way, by parliamentary committees such as the Standing Committees on Finance, Trade and Economic Affairs and the Standing Committee on External Affairs and National Defence (SCEAND). Government Members have the greatest influence (in caucus), partly because many interest groups are reluctant to approach opposition MPs whom they fear will use them to score political points and thus alienate the Government, and partly because ministers tend to pay less attention to opposition Members' interventions than to Government members. A former minister conveyed a candid piece of political horse sense in stating that "governments probably fear more losing a seat already won than hope to gain a seat by responding to opposition demands for protection."[17] The influence of opposition MPs does, of course, increase considerably in periods of minority government.

Parliamentary committees have proven a largely ineffectual avenue of backbench influence. The Finance, Trade and Economic Affairs Committee has discussed several pieces of trade legislation such as the Anti-Dumping Act in 1968, the Textile and Clothing Board Act in 1971, and amendments to the Export and Import Permits Act in 1974,[18] and periodically heard witnesses from the Cabinet and bureaucracy on the Multilateral Trade Negotiations. But most trade issues other than individual interest group demands are too secret and sometimes too technical to permit any real committee influence. One exception to this rule was the ability of the SCEAND Sub-Committee on International Development to liberalize somewhat Canadian positions at UNCTAD IV in 1976. This was accomplished more by the force of moral arguments relating to developing countries, to which the Prime Minister proved receptive, than by the technical expertise of Members of Parliament.

Somewhat in contrast to House of Commons Members, many Senators have exhibited a keen interest in the broader issues of trade policy. Senate committee reports have been prepared on Canada's trade and other relations with the Caribbean, the Pacific area, and the United States.[19] The 1978 Senate study on

Canada-U.S. relations proposed that serious consideration be given to establishment of a North American free trade zone. No doubt the business backgrounds of many Senators account for this greater awareness of broader trade policy issues. While the influence of the Senate is quite limited, as one might expect of a non-elected body, there has been the odd case of Senatorial influence. The Senate report on the textile industry, issued in Spring 1976[20], was, for instance, an important stage in the sequence of events leading up to global quotas on clothing in November 1976.

NOTES: Chapter Three

1. We have borrowed this notion from Donald W. Kelly, The Development of a New Textile Policy for Canada: A Case Study of Government-Industry Relations in Canada, unpublished D.B.A. Thesis, Harvard University, 1974.

2. S.D. Clark, The Canadian Manufacturers' Association (Toronto: University of Toronto Press, 1939).

3. Readers may recall the public debates over food policy during the hearings of Mrs. Beryl Plumptre's Food Prices Review Board, 1973-74. The role of imports in food policy sparked some heated exchanges between Mrs. Plumptre and the then federal Minister of Agriculture, Mr. Eugene Whelan.

4. Economic Council of Canada, Looking Outward: A New Trade Strategy for Canada (Ottawa: Information Canada, 1975) pp. 3, 32.

5. This argument has been made by several commentators, including the Science Council of Canada, Uncertain Prospects: The Canadian Manufacturing Industry 1971-77 (Ottawa: Minister of Supply and Services Canada, 1977).

6. The practice and implications of higher nominal and/or effective rates of protection for manufactured goods, are discussed in J. Melvin and B. Wilkinson, Effective Protection in the Canadian Economy, Economic Council of Canada, Special Study No. 9 (Ottawa, Queen's Printer, 1968) and B. Wilkinson and K. Norrie (Effective Protection and the Return to Capital, Economic Council of Canada (Ottawa: Information Canada, 1975).

7. Canadian Manufacturers' Association, Submission of the Canadian Manufacturers' Association to the Canadian Trade and Tariffs Committee with respect to the GATT Negotiations, Toronto, August 1974, p. 42.

8. *Ibid.*, p. 42.

9. Canadian Export Association, *CEA Submission to the Canadian Trade and Tariffs Committee*, Bulletin No. 306, 8 July 1975.

10. A few such books and survey articles on interest groups are: Robert Presthus, *Elite Accommodation in Canadian Politics* (Toronto: Macmillan, 1973); A. Paul Pross (ed.), *Pressure Group Behavior in Canadian Politics* (Toronto: McGraw-Hill Ryerson, 1975); Chapter 12 in R.J. Van Loon and M.S. Whittington, *The Canadian Political System*, 2nd ed. (Toronto: McGraw-Hill, 1976); W.T. Stanbury, "Lobbying and Interest Group Representation in the Legislative Process" in W.A.W. Neilson and J.C. Macpherson (eds.) *The Legislative Process in Canada: The Need for Reform* (Montreal: Institute for Research on Public Policy, 1978) Chapter 6; and Fred Thompson and W.T. Stanbury, "The Political Economy of Interest Groups in the Legislative Process in Canada," in Richard Schultz, Orest M. Kruhlak and John C. Terry (eds.), *The Canadian Political Process*, 3rd ed. (Toronto: Holt, Rinehart and Winston, 1979) pp. 225-249. The last two of these include up-to-date and quite comprehensive bibliographies.

11. S.D. Clark, *op. cit.*

12. Robert Presthus, *op. cit.*

13. John Kirton, *The Conduct and Coordination of Canadian Government Decision-Making Towards the United States*, unpublished doctoral dissertation, Johns Hopkins University, 1977, p. 84.

14. John Weir, "Trade and Resource Policies" in C. Winn and J. McMenemy (eds.), *Political Parties in Canada* (Toronto: McGraw-Hill Ryerson, 1976) p. 229.

15. *Ibid.*, p. 234.

16. One leading public figure, Robert Stanfield, has

called for a Canada-U.S. free trade arrangement. This was done after retirement. See Robert Stanfield, "Exploring a New Common Market," Globe and Mail, 8 February 1978. To date, no major party platform or active leader has publicly supported the concept. In the early months of the Clark administration, however, some journalists claimed to have uncovered sympathizers in the higher quarters of the Conservative Party. See, for example, Anthony Westell, "From Ottawa: Joe Clark's free-trade flirtation: A fundamental shift in Tory policy," Canadian Business (October 1979) p. 51.

17. Confidential interview.

18. See House of Commons, Standing Committee on Finance, Trade and Economic Affairs, Proceedings respecting the White Paper on Anti-Dumping, 1st Session, 28th Parliament 1968-69, Issues 2 (Oct. 24, 1968) to 17 (Dec. 5, 1968); Minutes of Proceedings respecting Bill C-215, "An Act to establish the Textile and Clothing Board and to make certain amendments to other Acts in consequence thereof," 3rd Session, 28th Parliament, 1970-71-72, Issues 15 (Feb. 2, 1971) to 23 (Mar. 2, 1971); Minutes of Proceedings and Evidence respecting Bill C-4, "An Act to amend the Export and Import Permits Act," 2nd Session, 29th Parliament 1974, Issue 10 (May 2, 1974).

19. See Senate of Canada, Report of the Standing Committee on Foreign Affairs on Canada-Caribbean Relations (Ottawa: Queen's Printer, 1970); Senate of Canada, Standing Senate Committee on Foreign Affairs, Report on Canadian Relations with the Countries of the Pacific Region (Ottawa: Queen's Printer, March 1972); and Senate of Canada, Standing Senate Committee on Foreign Affairs, Canada-United States Relations, Volume 2: Canada's Trade Relations with the United States (Ottawa: Queen's Printer, 1978).

20. See Senate of Canada, Proceedings of the Standing Senate Committee on Banking, Trade and Commerce: Canadian Textile Problems: Report of the Committee, Issue 82 (April 7, 1976).

Chapter Four

PROFILES OF TRADE POLICY ACTORS: THE SPECIALIZED REVIEW MECHANISMS

Review of competing interests' claims on government is the second stage of the typical trade decision-making sequence. Canada possesses no single body, such as the International Trade Commission in the United States, to perform this function. Instead, there are four different sets of review actors. In probably the majority of cases, the bureaucracy reviews the issues and merits of pressure group arguments. This is especially true of tariff policy where the great majority of reviews are conducted by the Tariffs Division of the Department of Finance, but somewhat less true of special import policy where the review function is shared between the Textile and Clothing Board and an Interdepartmental Committee on Low Cost Imports.

Over the years, three specialized review mechanisms have been created to make recommendations to government, each in a particular area of policy. We refer to the Tariff Board, the Anti-Dumping Tribunal, and the Textile and Clothing Board. These bodies were created for two major reasons. First, the General Agreement on Tariffs and Trade incorporated an international "norm" to the effect that special measures of protection should not be introduced without review by bodies independent of the executive branch of governments.[1] This norm was given formal status in Article 10 of the Agreement and later by the 1967 GATT Anti-Dumping Code and the 1974 Multifibre Agreement on textiles. It is clearly in the interests of all states that such independent bodies exist. The second reason, however, is related more to domestic politics. By referring to these bodies some of the most sensitive trade issues, ministers attempt to "take the heat off" themselves and disarm critics of whatever strip. Understandably, it is politically more comfortable for ministers either to resist or to accept demands for increased protection when the case has been investigated by an independent review body. Of course things do not always work out this way in practice; on numerous occasions, review body recommendations will come back to ministers for difficult political choice.

(1) The Tariff Board

Originally established in 1931, the Tariff Board was revived in 1949 after a period of wartime dormancy to meet Canada's obligations under Article 10 of the GATT. It has two principal functions: appeals and ministerial references. In its appeals function, the Board acts as a court of appeal in disputes between the Department of National Revenue and any parties, usually importers, who feel aggrieved by decisions of the Deputy Minister of National Revenue on customs valuation, tariff classification and determination of dumping. The appeals function is strictly one of interpreting the law as it stands. No recommendations are made as to what the tariff classification or valuation of goods should be.

The second, and from a policy perspective, more important function, is that of inquiries based upon ministerial references. Under Section 4.2 of the Tariff Board Act (R.S.C. 1970, T-1), the Minister of Finance can refer to the Board for inquiry "as to the effect that an increase or decrease of the existing rate of duty upon a given commodity might have upon industry or trade, and the extent to which the consumer is protected from exploitation." Such references amount to policy recommendations as to the proper rates and classification of duties, taking into account the interests of Canadian industries, consumers and international obligations. Over one hundred such reports have been issued to date, an average of about three or four per year, after public hearings at which are represented the views of producers, consumers and foreign interests.[2] Most of these cases have been ones of long-standing controversy where the competing interests have been quite vocal and the issues complex, and where it has been the opinion of Ministers that there should be a full ventilation of the issue via the adversary process employed in Tariff Board hearings. Board references have also been important to Ministers of Finance as means of attempting (not always successfully) to defuse difficult political problems. A classic case was the 1977 report on fruits and vegetables, a situation where the horticultural industry had been demanding increased protection for many years and where the importance of the food issue had been one on which consumer groups and foreign interests had been particularly vocal.[3]

The relevant interests seem on the whole satisfied with the independence and objectivity of the Board. Allegations of bias, which are regularly levelled against the Textile and Clothing Board by the Canadian Importers Association and the Consumers' Association of Canada, have been spared the Tariff Board. Producers are more concerned about the time it takes to file reports on appeals and references, but there are no fundamental criticisms. This relatively successful balancing would seem based on a tradition of appointments of impartial and regionally-balanced members to the Board, and the possession of an independent and highly qualified research staff. Perhaps the most unfortunate characteristic of the Tariff Board is that this independent body at present takes on only a small percentage of the tariff questions that arise.

(2) The Anti-Dumping Tribunal

Dumping, the act of selling goods in export markets for less than the home market price, is not necessarily bad. It can bring benefits to the importing country in terms of cheaper consumer goods and cheaper inputs into importing country manufactures. The GATT Anti-Dumping Code, negotiated during the Kennedy Round to supplement Article 6 of the Agreement, recognized that dumping is nefarious only if it causes material injury, or threat of material injury, to producers in the importing country, or if it retards the establishment of new industries in the importing country. Only under these circumstances should anti-dumping duties be applied. A new Canadian anti-dumping system was set up in 1969 to conform to the provisions of the Code. Previously the Canadian mechanism for dealing with dumped goods, based on Section 6 of the Customs Tariff, had received much international criticism, emanating primarily from Britain and the United States, for its lack of a formal, case-by-case test of material injury.[4]

The new system operates in two stages. First, the Department of National Revenue's Anti-Dumping Directorate determines whether dumping is occurring and by how much ("margin of dumping"), and applies an anti-dumping duty upon "preliminary determination" of dumping. The process normally begins with a complaint from Canadian firms, but the Deputy Minister of National Revenue also has the authority to initiate investigations of possible dumping on his own, or by reference from the Anti-Dumping Tribunal.

The second stage of the process involves an inquiry by the Anti-Dumping Tribunal, composed of four members and a chairman reporting to the Minister of Finance. The Tribunal conducts hearings at which all interested parties can be heard, to determine whether dumping has caused or threatened material injury to the production of like goods or is materially retarding the establishment of plants in Canada.

If no injury is found, the collected duties must be remitted. Even if injury is found, however, Cabinet may remit portions of the duty, for example to importers in certain regions of Canada, the market of which was not served by Canadian producers. An example of this occurred in early 1978 when pressure from British Columbia Members of Parliament succeeded in convincing Cabinet to remit duties to British Columbia importers of steel beams, because Algoma Steel Co., the injured Canadian producer, had not been a traditional supplier in British Columbia.[5]

Since its inception in 1969 to the end of 1976, the Tribunal reviewed 50 dumping questions. In 17 cases material injury or retardation of the establishment of plants was found, while in 7 likelihood of injury was found. The Tribunal decided that there was no injury in 21 cases, and in 5 other cases a combination of injury and no injury was found.[6] These statistics support the general impression amongst interest groups and policy-makers of complete objectivity on the part of the Tribunal.

The Anti-Dumping Tribunal has another important review role, one unrelated to dumping. Under Section 16.1 of the Anti-Dumping Act (R.S.C. 1970, A-15), the Tribunal is required to "inquire into and report to the Governor in Council on any other matter or thing in relation to the importation of goods into Canada that may cause or threaten injury to the production of any goods in Canada that the Governor in Council refers to the Tribunal for inquiry and report." This section was added to the Act in 1971 to cover petitions from Canadian manufacturers for protection from general import competition which could not be reviewed by the Textile and Clothing Board. To date only three Section 16.1 references have been undertaken by the Tribunal: canned mushrooms in 1973 (injury was found); footwear in 1973 (threat of injury) and again in 1977; this last inquiry led to global quotas on most footwear items in December 1977.[7] Recently (1977), the Anti-Dumping Tribunal was given the task of reviewing Revenue Canada's determination of subsidization in countervailing duty actions, but so far no such cases have arisen.

The major complaint of Canadian manufacturers against the new anti- dumping system involves the length of time, sometimes up to seven months, it takes to go through the process of preliminary determination of dumping and inquiry as to material injury, though this wrath is not particularly directed at the Tribunal. It is difficult, however, to see what can be done to speed up this process, given the technical difficulties involved and the nature of the GATT Code.

(3) The Textile and Clothing Board

The Textile and Clothing Board Act (S.C. 1970-71-72, 39) was passed by Parliament in May 1971 as part of a "Textile Policy," whose aim was to rationalize the Canadian textile and clothing industries through various instruments including adjustment assistance and temporary special protection. The role of the

Textile and Clothing Board (TCB) was to consider, by means of public inquiries, industry demands for protection, as a result either of representations directly to the Board or references to the Board from the Minister of Industry, Trade and Commerce. The three member Board is required to examine in confidence the adjustment plans of complaining firms. In rendering its judgement, it is obliged to take account of such factors as the consumer interest, Canadian obligations under the GATT and the Arrangement Regarding International Trade in Textiles, and, most fundamentally, "the principle that special measures of protection are not to be implemented for the purpose of encouraging the maintenance of lines of production that have no prospects of becoming competitive with foreign goods in the market of Canada if the only protection to be provided is that provided at any time by rates of duties of customs" (Section 18, TCB Act). The Act empowers the Board to recommend in an interim or final report "whether, in its opinion, special measures of protection should be implemented" (Section 17.(1)). These recommendations are then considered by the Minister of Industry, Trade and Commerce. The TCB is also obligated, under Section 19 of the Act, to "keep under constant review ... the progress of any adjustments made by producers of textile and clothing goods pursuant to plans submitted to the Board ... for the purpose of recommending to the Minister the modification or the removal of any specific measures of protection as soon as circumstances permit."

The Textile and Clothing Board has been surrounded in controversy since its establishment. This reflects the general political sensitivity of a troubled industry which employs a quarter million people, mostly in Quebec, and the importance attached to clothing by consumer groups, as well as the international development dimension of the problem. More specifically, questions have been raised about the Board's interpretation of the Act and the attitudes of its membership.

The first type of question frequently posed about the Board concerns its interpretation of Section 18 of the Act, quoted above. Given that protection over and above the tariff, which is in the 25 percent range on most textile items, is not to be provided to firms with "no prospects of becoming competitive," do wage rates count as an element of comparative advantage? The Canadian industry may be as technically efficient in terms of output per man as any in the world, but when all is said and done wage rates, and thus production costs, are far lower in developing countries (Korea, Hong Kong, China, Taiwan etc.) than in Canada. The Board has seemingly taken the view, increasingly in recent years as the possibilities for improving technical efficiency have become exhausted, that wage differentials are an illegitimate element of comparative advantage. Protection well beyond the tariff has consequently been offered the industry, in increasing doses (see Chapter Six). It is not surprising, therefore, that both the Board and the Textile Policy have come under sharp criticism from consumer groups, importers, academic economists, development interest groups and foreign governments, as well as in private from officials in the departments of Finance, External Affairs and even Industry, Trade and Commerce.[8]

The Board is also frequently accused of possessing a pro-industry bias, particularly by the Consumers' Association of Canada and the Canadian Importers Association, but also from some government sources. It has also been accused of being subject to political pressure. Critics charge that the Board has neither the capability nor the willingness to do an adequate job of keeping under constant review the progress made in industry adjustment plans. There seems to be merit in at least the capability side of the accusation. Several of these criticisms were summed up by a senior Industry, Trade and Commerce interviewee:

> You should understand that the Textile and Clothing Board was really created to give a political cushion to the Minister of Industry, Trade and Commerce, to get the Minister off the hook by providing him with sound advice that the interest groups could not criticize. Instead what has happened is that the Board just passes the can back to the Minister by continually

> supporting the industry point of view. They lack the resources for effective analysis of adjustment plans and confidentiality prevents them from being seen by we in Industry, Trade and Commerce, who have the expertise and have to make the decisions on protection.[9]

Even if these complaints are true, it may be improper to blame the Textile and Clothing Board. It is in a difficult position, caught in the crossfire between competing and vocal interests, and may have adopted its particular interpretation of Section 18 partly out of an instinct for self-preservation; one may well ask how long it would last if it interpreted wage differentials as a legitimate element of comparative advantage. The Board has not been given adequate resources to fulfill the "constant review" part of its mandate, and as long as the problem of Section 18 remains, it is not likely to be given these resources. The Act, moreover, is very restrictive in the powers of recommendation given to the Board, the only options being that "special measures of protection should be implemented" or should not. The Board has no power to propose, for example, packages of adjustment instruments, either in lieu of protection or in combination with temporary protection. As long as governments are not willing to accept the logical implications of Section 18 of the Act, the Board will continue to receive much of the criticism that might more properly be directed at the government.

In summary, we find a contrast between the Anti-Dumping Tribunal and the Tariff Board on the one hand, and the Textile and Clothing Board on the other, as to their ability to strike a reasonable balance between competing interests. The problems presented to the Textile and Clothing Board may well be more difficult than the average Tariff Board reference, though probably not much more difficult that certain Section 16.1 Anti- Dumping Tribunal references (e.g., footwear). Another reason for the Textile Board's lesser ability to keep all interests happy, or at least neutralized, lies in the greater precision of the law in the case of Tariff Board references and the anti-dumping process. These issues are less subject to political manipulation because there exists an international anti-dumping code, and Article 10 of the GATT. The international community has yet to agree on a precise interpretation of the "safeguards" in GATT Article 19, upon which the Canadian Textile and Clothing Board Act is based. Canada, in accordance with the traditional policy of reciprocity, is unlikely to relinquish unilaterally its ability to protect the politically-influential textile industry. This accounts in large measure for the controversies in which the Textile Board has found itself.

NOTES: Chapter Four

1. These bodies appear to be more "independent" than the other reviewers of trade problems - civil servants in executive departments, who are directly responsible to a minister. The Anti-Dumping Tribunal and the Tariff, and Textile and Clothing Boards may or may not be wholly independent of the political level. A recent report on regulatory and advisory agencies of this type raised the issue of whether ministers "influence ... regulatory agencies behind the scenes on the basis of political considerations - often in response to narrow interest group pressures." See Economic Council of Canada, Responsible Regulation (Ottawa: Minister of Supply and Services Canada, November 1979) pp. 53-68. The pages following report the belief of some interviewees that such attempts to influence the Textile and Clothing Board have been made. None argued this with respect to the Anti-Dumping Tribunal and Tariff Board. We of course have no evidence either to substantiate or to reject these opinions.

2. Two of the most important Tariff Board references during the Trudeau years concerned sugar and fruits and vegetables. The reports recommended reductions in sugar tariffs, and increases in duties on many fruits and vegetables. See Tariff Board, Report on Sugar, Ref. No. 146 (Ottawa: Information Canada, 1971) and Tariff Board, Report on Fresh and Processed Fruits and Vegetables, Volumes 1-5, Ref. No. 152 (Ottawa: Minister of Supply and Services Canada, 1977-78).

3. Ibid.

4. The definitive treatment of Canada's anti-dumping practices may be found in Rodney de C. Grey, The Development of the Canadian Anti-Dumping System (Montreal: Private Planning Association of Canada, 1973). See Chapter 2 for an account of the new 1969 system.

5. Ottawa Citizen, 31 January 1978.

6. These figures were compiled from Anti-Dumping Tribunal, *Annual Report 1976* (Ottawa: Minister of Supply and Services Canada, 1977) pp. 37-47.

7. See Anti-Dumping Tribunal *Report of the ADT under Section 16.1 of the Anti-Dumping Act Respecting the Effects of Preserved Mushroom Imports on Canadian Production of Like Goods* (Ottawa: Information Canada, 1973); Anti-Dumping Tribunal, *Report of the ADT under Section 16A of the Anti-Dumping Act Respecting the Effect of Footwear Imports on Canadian Production of Like Goods* (Ottawa: Information Canada, 1973); and Anti-Dumping Tribunal, *Report of the ADT under Section 16.1 of the Anti-Dumping Act Respecting the Effects of Imports on the Canadian Footwear Industry* (Ottawa: Minister of Supply and Services Canada, 1977).

8. Confidential interviews.

9. Confidential interviews.

Chapter Five

PROFILES OF TRADE POLICY ACTORS: GOVERNMENT DEPARTMENTS AND CENTRAL AGENCIES

It was observed earlier that the two principal roles of the bureaucracy in Canadian public policy-making are a review and analysis function and a recommendation function where civil servants attempt to work out agreed positions for recommendations to Cabinet. Trade policy is somewhat unusual in that formal responsibility for its instruments is fragmented among several ministries. Keen interest in trade issues is shown by an even wider array of departments and ministers, a fact which no doubt reflects the differential impact of trade decisions on socio-economic groups, regions of the country, macro-economic conditions, and relations with other countries.

In this chapter we shall analyse the roles and behaviour of departments and central agencies involved in making trade decisions over the period 1968 to 1979. The major aspects to be discussed are organizational power and responsibilities, relations with "client" groups, the value preferences of organizations, and their tactical behaviour over the period.

Doern has presented a useful classification of Canadian government portfolios by functions.[1] He divides ministries into four clusters, of which three are of interest here. The "traditional horizontal co-ordinative portfolios" have responsibilities which cut across many dimensions of government policy and include, of the departments and central agencies involved in trade policy, the Departments of Finance and External Affairs, the Privy Council Office, and the Prime Minister's Office. As Doern observes, these entities "have inherent high policy influence ... because they afford their occupants the highest number of strategic opportunities to intervene in almost any policy issue...."[2] Finance, moreover, has legal responsibility for several of the major instruments of trade policy. A second portfolio type is the "administrative coordinative" group which again cut across policy issues, but in a more technical way. The Department of National Revenue is the principal such actor on trade issues, although Supply and Services has a similar role on a few issues such as government procurement. A final portfolio category, numerically representing the largest number of departments, is the "vertical constituency" cluster. Their influence derives from their large program budgets and the fact that "they represent the vertical dimension of government in that they tend to extend outward to deal with their respective constitutencies."[3] The relevant departments for us will be Industry, Trade & Commerce, Agriculture, Consumer and Corporate Affairs, Energy Mines & Resources, Employment and Immigration, Regional Economic Expansion, Fisheries, Labour, and the Canadian International Development Agency. Industry, Trade and Commerce is more difficult to classify than the other departments. Its industry side is fairly clearly of the "vertical constituency," or representative type. Its trade and commerce component could be put in the horizontal category because of its lesser formal attachment to particular groups in society, but some observers do see its mandate of expanding exports as entailing a client relationship with Canadian exporters.

It is well established that there exist significant value differences between government departments, and sometimes between sub-units of departments, on policy issues. Trade policy is no exception. The recent Canadian International Image Study, involving interviews with some 300 senior officials and ministers on trade and other foreign policy questions, revealed this strikingly.[4] The analysis to follow points to three major sources of trade value cleavages between departments and sub-units: the general role of the organization in the government system as defined by Doern's portfolio types; the functions of sub-units in the trade policy process (e.g., units whose main function is to negotiate the removal of foreign barriers to Canadian exports versus units administering state aid to domestic constituencies); and finally the views and interests of incumbent ministers and deputy ministers.

(1) The Department of Industry, Trade and Commerce

The trade policy role of the Department of Industry, Trade and Commerce and its predecessors has varied somewhat over the years since World War II. During the C.D. Howe era, Trade and Commerce was dominant, and even impinged on the more general macroeconomic policy role of the Department of Finance.[5] Through the Diefenbaker and Pearson periods, Finance appears in greater control of trade policy, in large measure due to the importance of tariff negotiations in the Kennedy Round and the appointment of particularly strong Ministers and Deputies. Rough equality between Finance and IT&C emerged during most of the Trudeau era when IT&C regained much authority as a result of the new importance of non-tariff barriers and commodity questions in international trade relations, and the efforts of the Trudeau government to integrate trade and industrial development policy. At the end of the Trudeau decade, there were indications that relative power was once again shifting away from IT&C - the indications being the appointment of a Minister (Mr. Jack Horner), a former Conservative preoccupied with his personal problems of re-election, the appointment of the Deputy Prime Minister (Mr. Allan MacEachen) as chief Cabinet coordinator of the Multilateral Trade Negotiations, and the creation of a Board of Economic Development Ministers in 1978 under a Minister (Mr. Robert Andras) other than the IT&C Minister. It was not clear in early 1979, however, that the direct beneficiary of this apparent decline would be Finance. These remarks concern, of course, only broad trends, as trade decision-making has been a largely shared activity between (at least) Finance, IT&C and External Affairs.

The legal authority for IT&C's key role in trade policy is derived from two statutes, the Export and Import Permits Act and the Industry, Trade and Commerce Act. These give the Minister primary responsiblity for special import policy, or the imposition of quantitative restrictions on imports and exports, export promotion, the conduct of international trade negotiations, grains sales and the Canadian Wheat Board, and international commodities policy. Given its expertise on industry sectors, IT&C's influence extends into tariff policy as well. The trade role of the department is further augmented through its responsibiltiy for general industrial policy and programs of assistance and incentives to Canadian industries.[6]

For most of its existence since 1969, Industry, Trade & Commerce has been divided into five major branches, of which two directly concern us. The Industry Development Branch, under an assistant deputy minister, includes the "sector divisions" which monitor the performance of, and undertake liaison with, the chemicals, agriculture, resources, textiles and consumer products, electronics, and machinery sectors of the economy. The department's "vertical constituency" or representative role is most evident in these divisions, although their trade interests (i.e. protectionist vs. liberal) are not at all identical.

International Trade Relations Branch, the unit most directly responsible for trade instruments, has three components. Its Office of General Relations is concerned with developing Canadian postures in the multilateral trade institutions (GATT, UNCTAD, OECD), negotiating international commodity agreements, and generally pursuing Canadian export interests in both multilateral and bilateral contexts. The Office of Special Import Policy has the primary responsibility for policy and implementation of quantitative import controls. The International Bureaux, which joined the Branch in 1977, are mainly concerned with bilateral trade relations and export promotion.

During most of the period 1968-1979, a fairly clear clash of priorities existed on many issues between the International Trade Relations Branch (particularly the Office of General Relations) and the sector or line divisions of the Industry Development Branch. The roots of this conflict went back to the amalgamation in 1969 of two departments to create IT&C. The old Trade and Commerce Department had been in existence since the late 19th century. A Department of Industry was created in 1963 as a "vertical constituency" department, largely as a result of representations from Canadian manufacturers who maintained that, since farmers had the Department of Agriculture and fishermen the Department of Fisheries, they should have "their" department. The Industry department recruited its staff primarily from private industry, and this practice continued to a significant (though gradually declining) extent into the seventies in the sector divisions of the new IT&C department.

A major motive for the merger of the two portfolios was to integrate planning and implementation of two mutually reinforcing themes, domestic industrial rationalization (incentives to promote efficiency and innovation) and export expansion (trade promotion and negotiations to reduce foreign import barriers). Planning and integration were, or course, important concepts in the thinking of the new Trudeau team of 1968, particularly of Michael Pitfield, Clerk of the Privy Council and a major architect of amalgamation of the two departments.[7]

Unity of purpose, however, became more difficult to achieve on many practical questions than on organizational charts. While structural integration proved largely successful in relation to industries which were already competitive in international markets, great difficulties were experienced for several years in coordinating the purposes of those sector divisions dealing with "problem" industries on the one hand and the International Trade Relations Branch (ITR) on the other. Given that ITR's major function is to facilitate Canadian exports by negotiating the reduction of foreign barriers to Canadian exports (through the GATT rounds and bilateral initiatives on such issues as American countervailing duties), it tends to be wary of new Canadian barriers which may undermine such negotiations. This, however, was precisely what many of the "weak" sector divisions were proposing, particularly in fields like textiles, footwear, electronics and many agricultural products. While it may be somewhat simplistic to label these divisions as "protectionist," particularly since the protective measures proposed were ostensibly designed to provide temporary relief to firms during a period of rationalization, positions taken by the inward-looking sector groups were frequently at odds with those of the trade policy branch of IT&C. Conflicting regional pressures on Ministers also contributed to discord.

Some evidence suggests a narrowing of the value differences between the two branches in the last two or three years. The line divisions are beginning to assume a more "directionist" posture toward their sectors, in some contrast to the representational focus of their earlier role. A series of 23 sector studies were initiated in 1976 to plan government goals with respect to particular industries and to evolve packages of industrial and commercial policy incentives aimed at furthering international competitiveness.[8] This more positive approach to government-industry relations was illustrated by the television sector strategy, which allowed for temporary protection from imports during a period of rationalization in which manufacturers would shift production from TV sets to other home entertainment products in which Canada stood a chance of being competitive. No doubt, the Multilateral Trade Negotiations contributed to this policy re-orientation by opening the prospect of a far more competitive world trading environment, as did the decreasing tendency of the Department to staff its sector divisions with former employees of the industries. At the same time, International Trade Relations Branch has become more tolerant of temporary protective measures grounded firmly in a rationalization strategy. This narrowing of the gap between the line branch and the trade policy branch is a recent development, however, and did not apply during most of the period covered by this research.

The degree to which conflicts within IT&C were resolved, and the form that resolution took, depended greatly on the personal priorities and biases of Ministers and Deputy Ministers, current thinking in the interdepartmental setting, and prevailing economic and political circumstances. Ministers such as Jean Chrétien and Jean-Luc Pépin tended to sympathize with the weak sector divisions and were prepared on occasion to override advice emanating from ITR Branch, probably because they were Quebec ministers from constituencies where many of the most beleaguered industries are located. Alistair Gillespie and Jack Horner are generally thought to have favoured the trade policy orientation (Horner) or been relatively neutral (Gillespie).[9] The senior bureaucrats, in any case, are in the habit of anticipating the biases of their Ministers and presenting recommendations accordingly. The sector divisions were thus less likely to offer recommendations of the more extremely protectionist kind to Ministers like Horner and Gillespie than they did to Chrétien. The biases of Deputy Ministers were also relevant in this regard, probably never so much as during the Horner period when the Minister was preoccupied with the danger of personal defeat in the next election.[10]

An attempt was made to reconcile the differing perspectives of the two "departments" within IT&C prior to referring the matter to interdepartmental discussion. This was not always achieved, in a few cases a minority opinion having been expressed in draft memoranda to Cabinet, and in some cases the conflicting preferences resurfaced once the question went to interdepartmental committees.

(2) The Department of Finance

When all policy fields are considered, Finance is probably the single most powerful department in the Government of Canada. As an "horizontal coordinative portfolio," its roles include general management of macro-economic policy via the budget, and the central agency roles of controlling the spending plans of the program-oriented departments and coordinating the activities of these departments into a coherent economic policy. Along with IT&C and External Affairs, Finance is invariably involved in all trade decisions. Its key role derives not only from its central position within the government, but also from direct statutory responsibility for several instruments of trade policy, including tariff setting under the Customs Tariff, setting policy (though not administration) under the Customs Act and the Anti-Dumping Act, and references to the Tariff Board. Finance's formal responsibilities in trade matters are fewer today than they were during the 1960s, before the primary legal mandate for international trade negotiations and quantitative restrictions was transferred to IT&C. But preparations for the Multilateral Trade Negotiations and the making of special import policy have been interdepartmental exercises to such a degree that Finance's loss of power in these areas is more _dejure_ than _de facto_.

The focal point of trade activity within the department is presently the International Trade and Finance (ITF) Branch (corresponding roughly to the Tariffs, Trade and Aid Branch before 1974), headed by an Assistant Deputy Minister. The ITF Branch is divided into the Tariffs Division, the International Economic Relations Division, the International Finance Division, and the International Programs Division. International Finance Division, the locus of international monetary and balance of payments policy, plays only an indirect role in trade policy, but issues with a major effect on the balance of payments (such as the 1971 American surcharge and the Auto Pact) are closely followed by this division. The International Programs Division, whose main responsibilities are foreign aid policy and relations with UNCTAD, has had a somewhat greater direct role in recent years because of the preoccupation in UNCTAD with commodity trade problems.

Tariffs Division is of course the unit primarily responsible for tariff setting, not only in terms of evaluating interest group representations on customs duties, but also by preparing the tariff component of the Government's budget and elaborating Canadian tariff offers at international negotiations. The division also prepares and reacts to references to the Tariff Board, participates in policy on the _Customs Act_, and is consulted about non-tariff negotiations because of the interrelationships and substitutability of tariff and non-tariff measures of protection.

The International Economic Relations (IER) Division corresponds roughly in both function and orientation to IT&C's Office of General Relations. As the division of Finance most directly concerned with furthering Canadian export interests (i.e., removing foreign barriers), its approach is somewhat more liberal than that of Tariffs Division, which finds itself in daily contact with Canadian interest groups seeking protection and is dependent to some degree for information on IT&C's industrial sector branches. IER is in addition the focal point within Finance for special import policy, although there is close consultation with the Economic Development Division or industrial policy wing of Economic Programs and Government Finance Branch. On commodities policy, IER shares responsibility with International Programs Division (IPD), with IER tending to concentrate on individual commodity agreements and IPD on the "aid" aspects such as UNCTAD's Common Fund plan. Finally, it is IER which develops policy for the Finance Department on the anti-dumping system, countervailing duty regulations, and on codes respecting non-tariff barriers at multilateral trade negotiations.

Despite modest differences of orientation between Tariffs Division and International Economic Relations Division, the Department of Finance has spoken with one voice on the interdepartmental scene to a greater degree than Industry, Trade and Commerce. There are two major reasons. First, the relatively less liberal unit, Tariffs Division, is in intensity less protectionist than the Industry side of IT&C was for the most of the period 1968-79, partly because it is not as closely related to client interest groups as any given sector branch of IT&C. When an issue comes up relating to say, bicycles, it is the Textile and Consumer Products Branch of IT&C which will appear on interdepartmental committees, and this branch has close and sympathetic ties to that industry. Tariffs Division of Finance is not organized in such a sectoral way (though it naturally has its industry specialists) and thus more independence is possible. Second, unlike corresponding actors in the Department of Industry, Trade and Comerce, Tariffs and IER of Finance share the same Assistant Deputy Minister and thus divisions of opinion are more likely to be resolved before issues move into the interdepartmental phase. For example, the Assistant Deputy Minister directed his General Director [sic] in early 1978 to coordinate all Finance activities relating to the Multilateral Trade Negotiations, and this encouraged internal resolution of conflicts.

The general trade policy values of the Department of Finance are difficult to classify as either protectionist or liberal, in contrast to most departments of the "vertical constituency" variety, and even another horizontal organization like External Affairs. Broadly speaking, Finance exhibits less instinctive bias, and more pragmatic behaviour, than other departments. Several reasons seem determinant. First, its role as a central agency of government ensures that the Department is not tied to particular interests within society. The lack of sectoral organization in Tariffs Division is indicative of this. There is also a strong tendency among Finance officials to view themselves as the grand synthesisers of the biases inherent in any trade problem, and this self-image no doubt influences their behaviour. Second, because of its role as manager of the economy as a whole, Finance tends more than other departments to attempt an integration of trade policy with current trends in general economic policy. This may lead to either protectionist or liberal positions, and sometimes to stand-pat positions. In 1969 and 1973, for example, the general thrust of macro-economic policy was to reduce inflation; broad cuts in tariffs were incorporated into budgets. In periods of greater unemployment, Finance has been more accepting of protective measures. A third reason for Finance's relative value neutrality relates to the constraints of the GATT and the negotiating realities of that slice of trade policy most firmly in Finance control, tariffs. Raising tariffs beyond their statutory levels is subject to retaliation and equivalent compensation under GATT Article 28. Unilateral lowering of duties (except on a temporary basis) is seen by Finance as a giveaway of bargaining power that it might need in future negotiations (and accounts for Finance's hostility to the Economic Council's 1975 proposal to do just this).[11]

The result of these considerations is that, in relation to other departments, Finance follows a kind of trade policy orthodoxy which amounts basically to resistance to unilateral lowering and raising of import barriers, but sympathy toward liberalizations negotiated on a reciprocal basis. Depending on one's point of view, this general attitude may be regarded as realistic, unimaginative, or, as is the tendency among constituency-oriented departments, a frustrating negativism toward concerns that these departments value highly.

In addition to current macro-economic policy, the trade positions of the Department of Finance greatly reflect the views of incumbent Ministers, Deputy Ministers and, uniquely, Parliamentary Secretaries to Ministers. Like their colleagues in other departments, trade officials in Finance attempt to anticipate the bent of their Deputy and their Minister. As a rule, Deputy Ministers of Finance have delegated most trade issues to the Assistant Deputy Minister for International Trade and Finance, though this was less true of Simon Reisman, who had a keener personal interest in trade than his predecessor Robert B. Bryce or his successor T.K. Shoyama. The Department of Finance has been unique in that a tradition has been built up in the past decade or so, begun by Mitchell Sharp, of using the Minister's Parliamentary Secretary as an understudy or "junior minister in training." These secretaries have tended to specialize in taxes and tariffs, guiding legislation through the House of Commons, and have had significant involvement in internal Finance tariff decisions. Three of the four Finance Ministers in the Trudeau Government (Chrétien, MacDonald and Benson) were at one point in their careers Parliamentary Secretary to the Minister of Finance.

The Minister of Finance, of course, sets the tone for trade policy, and his known views and political interests are taken into account by his officials in their recommendations. Prime Ministers of Canada have on the whole not interferred greatly with their Ministers of Finance. The personal styles and biases of Ministers of Finance have affected tariff policy probably more than general macro-economic policies, given that trade policy is more viscerally understandable to politicians than macro-economic policy. Several knowledgeable people believe, for example, that had the ardent nationalist Walter Gordon remained as Minister until 1967, Canada probably would not have so easily accepted the substantial Kennedy Round tariff reductions that it did under Mitchell Sharp.[12]

(3) The Department of External Affairs

The Department of External Affairs (DEA) has always been one of the three key departments on trade issues, and its substantive role has increased somewhat in the past decade. The legal basis of this involvement is the Department of External Affairs Act (R.S.C. 1970, E-20), which gives the Secretary of State responsibility for the "conduct and management of international negotiations." Several orders-in-council have given DEA the primary responsibility for organization of international conferences, international treaties, chairmanship of several interdepartmental committees on trade matters, including the interdepartmental committees on Commercial Policy (ICCP), Economic Relations with Developing Countries (ICERDC), the first MTN committee, and for the coordination of all Canadian foreign operations and interactions.

The focal point of External's role in the trade policy-making process is the Bureau of Commercial and Commodity Relations, headed by a Director-General, and composed of two divisions, the Commercial and General Economic Policy Division and the Commodity and Energy Policy Division. This structure was based on a 1977 reorganization of the Department. Between 1971 and 1977 the relevant division was the Commercial Policy Division.

External's capabilities in trade matters traditionally rested on expertise in international law (including international economic law), provision of intelligence from the missions about the intentions and potential reactions of foreign governments, and the prestige of the Secretary of State and a few External mandarins, such as A.E. Ritchie and Norman Robertson. During the 1950s and 1960s, the Department did not compete with Finance and Trade and Commerce in conducting economic analysis. Its trade policy role revolved around supplying intelligence and legal inputs, coordinating trade and foreign policy objectives, and mediating disputes between Ottawa departments, including Finance and Trade and Commerce. These roles amounted to substantial influence for the Department.

The role crisis experienced by External Affairs beginning in the late 1960s has been well documented by Thordarson.[13] Malaise about the Department's role in trade policy and other fields was traced to several disparaging remarks by Prime Minister Trudeau about the obsolete role of diplomats, Trudeau's disinclination to follow the custom of placing his most prestigious minister and heir apparent in the External portfolio, the relative lowering of its Undersecretaries on the Ottawa totem-pole of deputy ministers, improvements in international communications which diminished External's intelligence inputs, and the increasing complexity of international economic matters which presented problems to a department composed of generalist officers. Faced with these challenges, External attempted to carve out new roles for itself in foreign policy generally and these affected its trade policy activities. These roles emphasized an upgrading of internal economic expertise and a more explicit conception of the Department as a coordinating "central agency" of government.

A concerted effort was made to strengthen External's purely economic expertise by allowing occupational specialization of officers in economic fields and subjecting them to less of the usual rotation to which most generalist officers are subject. The 1977 reorganization of the Department also created a new position of Deputy Undersecretary for Economic Relations, which was intended to upgrade DEA's trade and other economic representation in interdepartmental dealings. This new economic thrust was initiated after the foreign policy review of 1970 had emphasized "economic growth" as the primary theme of Canadian foreign policy.[14] Events unfolding during the 1970s accentuated the trend as economic problems, like monetary and energy crises, the "new economic order" and the multilateral trade negotiations came to the fore in international affairs.

Yet it would be gross exaggeration to equate External today with Finance and IT&C in expertise on the substantive economic issues of trade policy. DEA does contribute more now to the review process and economic analysis, and its interventions are coming at an earlier stage of the policy development process. But the result of the post-1970 upgrading of economic resources in the department has essentially been one of narrowing the gap vis-à-vis IT&C and Finance just enough to permit External to better play its traditional coordination role in the more complex environment of the 1970s.

This role as a "horizontal coordinative portfolio" stresses, in the first instance, consistency of trade policy outputs. Efforts are made to ensure that what Canada does in an arena like UNCTAD is not in contradiction to what it is doing under the GATT. While such efforts are desirable, their success is far from complete. Furthermore, External performs much of the less than glorious footwork necessary to coordinate departmental inputs into various conferences such as the Multilateral Trade Negotiations, and its control of the international communications system (telexes etc.) is a source of some influence. But probably the most valuable coordination role played by External is the continuous attempt to achieve consistency between trade policy outputs and other foreign policy objectives. The most visible examples in recent years have come in the areas of international development and the multi-thematic North-South dialogue, Western heads of government summits, and moral issues such as those raised by trade with South Africa. External's striving for consistency has at best, and probably inevitably, been only imperfectly realized. It is also a frequent source of differences of view between DEA and Finance and IT&C, never more dramatically demonstrated than in External's unsuccessful proposal in 1970-71 to integrate the IT&C trade commissioner service into a unified foreign service.[15]

DEA's concern for coordination, and the lesser interest of Finance and IT&C in this function, was illustrated by the story of the Interdepartmental Committee on Commercial Policy (ICCP). Inspired and chaired by External, this committee was set up in 1972 at the assistant deputy minister level to provide a forum for discussion of the broad issues and ends of commercial policy. It includes most of the departments with an interest in trade policy and DEA had hoped that ICCP would prove a valuable mechanism for briefing the secondary departments, as well as for communicating the concerns of these departments to Finance and IT&C. The two heavyweight trade departments, however, have never shown much enthusiasm for such exercises, preferring instead meetings on highly specific topics with a restriced number of departments represented. IT&C and Finance, moreover, are little disposed to engaging in "philosophical" exercises with departments whose expertise is in their eyes wanting. Largely as a result of this lack of interest, the frequency of meetings of ICCP, and the level of representation, has declined over the years.

External Affairs' trade "ideology" favours respect for international obligations under the GATT, sensitivity to Canadian export interests, a desire not to upset foreign governments through protectionist actions, and the taking account of diplomatic or foreign policy aims in trade decisions (e.g., international development and the "third option").[16] Like the Department of Finance, External Affairs has no clear constituency linkages but, for the above reasons, External appears more consistently liberal in orientation than the more neutral Finance. This is most apparent on issues relating to import problems posed by developing countries. But DEA's liberalism is more moderate than that of departments like Consumer and Corporate Affairs (see below).

(4) The Department of Agriculture

Among "secondary" trade policy departments in Ottawa, it is the Department of Agriculture which over the years has enjoyed the most consistent involvement in trade issues of interest to it. This unique role is based not on statutory authority for trade instruments, but rather on the exceptionally strong political power of its constituency, the Department's recognized expertise on farm issues, the fact that price support and marketing arrangements in the agricultural sector must be supplemented by measures relating to imports if they are to be effective, and a long tradition of consultation by Finance and IT&C with this oldest of "vertical constituency" departments.

While it is clearly the objective of the Department to promote the interests of the farm sector, Agriculture's positions inevitably reflect choices and trade-offs between the divergent interests of generally export-oriented Western Canadian farmers and import-sensitive Eastern farmers. The Western grain producers' objective of greater access to European and Japanese markets, for example, is difficult to accomplish without making concessions to these countries in sectors in which they possess cost and quality advantages. This runs headlong into the vested interests of Eastern dairy, poultry, vegetable and wine producers who depend greatly on tariffs and the Import Control List. In recent years, Canada has been fortunate in avoiding the most painful choices posed by this conflict because of healthy grain sales to the U.S.S.R. and the Peoples Republic of China, but the dilemma reared its head in the Geneva negotiations, and on several other issues. During the Trudeau era, the appointment of primarily Eastern Ministers of Agriculture, in particular the long-lasting Eugene Whelan, as well as the importance of the Ontario rural vote to the Liberal party, generally gave the edge to the Eastern farm interests. It was largely the result of Mr. Whelan's intervention that the <u>Export and Import Permits Act</u> was amended in 1972 to allow additions to the Import Control List based on "national supply management schemes." It was pressure from the horticultural sector communicated through Whelan to the Minister of Finance that accounted for a major Tariff Board reference on fruits and vegetables.[17] It resulted in higher tariffs on these products in 1978-79.

These East-West differences have their administrative counterpart in the differing trade perspectives of the Trade Policy Section of the Marketing and Trade Division, and the Department's Production and Marketing Branch. The Trade Policy Section is in effect the "external office" of the Department and its responsibilities include representation on interdepartmental committees and the coordination of departmental initiatives in multilateral trade negotiations. It tends to emphasize export opportunities and interests. By contrast, the Production and Marketing Branch, whose primary function is the administration of the various marketing statutes, is divided into divisions liaising with each agricultural sector. The Divisions tend to represent vertically the narrow interests of the sectors (e.g., Dairy Division, Fruit and Vegetable Division, Grains Division), many of which are protectionist in orientation.

The role of the Department of Agriculture in trade policy-making, always greater than that of other secondary departments, has further increased in recent years, largely because rising intervention in domestic markets by governments has entailed greater attention to imports as a complement to marketing and price support policies. The increasingly vocal consumer movement and the controversies during the mid- 1970s between the Minister of Agriculture and Mrs. Plumptre's Food Prices Review Board[18] have also focussed greater attention on trade matters by the Department, not to mention the fact that the trade negotiations in Geneva reopened the prickly subject of barriers to trade in agricultural commodities.

On tariff issues, Agriculture is virtually always closely consulted by the Department of Finance. This applies to both ongoing tariff policy via the budget, and to the preparation of the Canadian offer and request lists for GATT rounds. Consultation is equally close with IT&C on the Import Control List and the MTN codes on standards, technical barriers and subsidies/countervail. On these issues the Department of Agriculture comes regularly into conflict with Consumer and Corporate Affairs, and periodically with the Departments of Finance and External Affairs. The latter sometimes worries about the addition of more "irritants" with the U.S. when items are put under quantitative or tariff restraint. On commodity issues, Agriculture tends to react favourably to international commodity agreements to smooth out price fluctuations in wheat and sugar markets and would like to see new agreements for meat and dairy products. There is some conflict with the more free market orientation of the International Trade Relations Branch of IT&C, but close consultation exists nonetheless.

(5) Department of National Revenue (Customs and Excise)

The basic trade function of the Department of National Revenue (Customs and Excise) or Revenue Canada is to administer, at ports of entry, the provisions of the Customs Act (on value for duty), the Customs Tariff (on tariff rates and classification), the Anti-Dumping Act (determination of dumping and margins of dumping) and the Countervailing Duty Regulations. The department also enforces about 80 federal and provincial statutes and regulations as they apply to imports, including the Agricultural Products Standards Act, the Food and Drugs Act, the Consumer Packaging and Labelling Act, the Hazardous Products Act, and the Textile Labelling Act.

The classification of National Revenue as an "administrative coordinative portfolio" is certainly valid with regard to trade, but the department occasionally performs an important policy role. One such policy function involves the collection and analysis of data for decision-making purposes (for example, to help determine whether imports are in fact causing injury to Canadian producers) and the provision of technical information for evaluating the feasibility of various options. The department plays this informational role in several interdepartmental committees, including those on Low Cost Imports, the Generalized System of Preferences (GSP), and the Multilateral Trade Negotiations.

National Revenue also contributes to the drafting stages of legislation affecting imports. This role was illustrated by the 1977-78 review of the Customs Act, prodded by complaints by foreign governments that the Act's value-for duty and other provisions were obsolete and protectionist in intent. The review consisted of consultations with the Departments of Finance and IT&C, as well as several interest groups.

Information gathered served as input into Canada's negotiating hand in Geneva, and revisions to the Customs Act were linked to concessions made by trade partners.

A third policy role of Revenue Canada is the variable administration of customs and anti-dumping regulations according to current economic conditions. Procedures are generally tightened up to provide maximum protection within the law during periods of high import pressure on domestic producers and loosened in times of economic boom. For example, after the Anti-Dumping Tribunal determined in 1973 that imports were not causing injury to Canadian footwear companies, value-for-duty investigations were undertaken by Revenue officials in Italy and Spain, with the result that valuations were significantly increased on certain shoes. It is not clear to what extent such actions are the result of Cabinet deliberations, but it is certainly plausible that ministers under pressure from domestic interests would be tempted to initiate them.

On balance, though, Revenue Canada's policy role is quite modest. Most Ministers and officials of the Department have seen their function primarily in terms of implementation and have not sought any great influence over policy. Indeed, the major reason for Revenue's participation in interdepartmental committees, such as those on Low Cost Imports and the GSP, is that virtually all conceivably relevant departments were given membership after Consumer Affairs and CIDA clamoured for an input. A second reason for this modest role is that, even on those issues where Revenue would have wished a larger policy role, its Cabinet representation has been weak. There were no less than nine Ministers of National Revenue during the 1968-79 period, most of whom stayed for about one year and were of junior rank in the Cabinet hierarchy, at least at the time of their incumbency at Revenue. They were Chrétien, Côté, Gray, Stanbury, Basford, Cullen, Bégin, Guay and Abbott.

(6) The Department of Consumer and Corporate Affairs

The Consumer and Corporate Affairs Act of 1967 (R.S.C. 1970, C-27) gave this department (CCA) the twin mandate of initiating "programs designed to promote the interests of the Canadian consumer" (Section 6.1) and ensuring maximum competition in the marketplace by means of various regulatory instruments, policy on corporate mergers, and structrual re-organization of markets to encourage efficiency. CCA officials have construed the spirit of these mandates as legitimizing a strong role for the Department in trade decision-making, but this interpretation has been only partially shared by the key trade policy departments.

CCA's inputs into trade policy issues are centralized in its Coordination Branch, which represents the Department on interdepartmental committees and consults with the Bureau of Competition Policy's sector divisions.

Although responsible for a number of regulatory consumer-protection statutes (such as the Consumer Packaging and Labelling Act, the Textile Labelling Act and the Hazardous Products Act) which other countries have identified on occasion as non-tariff barriers to trade, CCA's views on most trade issues are highly liberal. The Department was quite deliberately created in 1967 as a "vertical constituency portfolio" designed to promote the consumer interest in the federal Cabinet. By extension on trade matters, this representation has come to include the interests of importers, and both the Consumers' Association of Canada and the Canadian Importers Association have been included in the Canadian Consumer Council, an official advisory body to the CCA Minister.

CCA's trade views have brought it into frequent disagreement, not only with the protection-minded departments, but also, on a more conceptual plane, with the relatively neutral Department of Finance. Generally speaking, Finance follows the orthodox GATT norm of avoiding protection, but will recommend it if injury to Canadian firms from imports can be clearly demonstrated. Because of its mandate for promoting structural conditions which favour maximum competition in the marketplace, Consumer and Corporate Affairs basically rejects this injury-safeguards reasoning and proposes contraction and restructuring of injured industries via (among other measures) a liberal commercial policy. Again, Finance takes the orthodox view that such uses of commercial policy, while desirable in the long-run, should only be undertaken through multilateral negotiation based on reciprocity, and not through unilateral liberalization.

The result of these differing perspectives is that Finance officials tend to regard CCA officials as politically naive and dogmatic. CCA tends to see Finance as overly concerned with the short-term and unimaginative in its thinking about the ends of trade policy. The disagreement has probably accounted in some measure for a certain disinclination on Finance's part to consult regularly with Consumer and Corporate Affairs on tariff issues that arise between GATT rounds. Such tariff consultation is largely _ad hoc_ and limited to issues where the Minister of CCA has recieved specific representations from interest groups and to the General Preferential Tariff, where diplomatic considerations and central agency pressures basically forced a full-scale interdepartmental committee on a reluctant Finance. Of course, the limits to CCA's trade resources have also contributed to a lesser involvement in trade issues than its mandate would suggest as appropriate, particularly in the case of the Multilateral Trade Negotiations, where CCA officials effectively restricted themselves to preparing their Minister for participation in the special Cabinet Committee on the MTN.

CCA has played a more regular role in special import policy, partly because of greater receptivity to its participation on the part of the International Trade Relations Branch of IT&C, and also because goods faced with the prospect of addition to the Import Control List tend to be very high profile consumer products. CCA participates in the Interdepartmental Committee on Low Cost Imports insofar as manufactured goods from developing countries are concerned, and is regularly consulted on an informal basis concerning agricultural Import Control List issues. Its position in these consultations has been the rather straightforward one of resisting additions to the List.

The influence of the department on trade decisions, however, has proven disappointing to its officials. No doubt this reflects the relative strength in our society of producer as compared with consumer interests, as well as the rapid turnover of Ministers (eight in the Trudeau government, mostly of limited weight in the Cabinet) and of Deputy Ministers (six during the same years). An additional factor is probably that the very predictable anti-protectionist stands of the Department tend to invite less consultation by the Department of Finance than might otherwise be the case.

(7) The Canadian International Development Agency

Although CIDA was established in 1968 as a government department reporting to the Secretary of State for External Affairs, it was not until 1974-75 that the Agency became more or less accepted as a member of the interdepartmental community on several important trade issues, including the General Preferential Tariff, international commodity negotiations, special import policy and the proposal to create a Trade Facilitation Office in Canada for developing countries. Prior to 1974, the Departments of Finance and IT&C doubted the need for such involvement by an aid agency with no statutory trade responsibilities, holding the view that CIDA could contribute nothing that External Affairs did not already contribute. Within CIDA itself, moreover, the Agency's role was largely perceived as limited to the delivery of foreign aid, and consequently its involvement in trade issues before 1974 amounted to *ad hoc* participation in a few Tariff Board studies and representation at UNCTAD meetings.

CIDA's trade role became more extensive and regular after the concept of a "new international economic order" (NIEO) came to the fore in international relations in 1974. The NIEO, as proposed by developing countries in the United Nations, seeks to reform economic relationships between rich and poor countries on several fronts - trade, technology, finance and private investment, as well as through the traditional aid approach to assisting the Third World. CIDA officials began actively to seek out a role in the interdepartmental discussion of trade issues as they related to developing countries. Finance and IT&C remained lukewarm to the idea, but came to accept full CIDA membership on several interdepartmental committees at the insistance of External and the Prime Minister's Office.

Until recently, research on trade issues and participation in committee work was shared between CIDA's Policy Branch and the Policy Institutions Division of the Multilateral Branch. Policy concentrated on the General Preferential Tariff and Low Cost Imports Committees, while Multilateral had the main responsibility for commodities and other UNCTAD issues. In early 1978, as a result of a full-scale reorganization of the Agency, the two divisions were amalgamated into a new Cooperation and Coordination Division in Policy Branch which would handle all non-aid issues on behalf of CIDA.

Throughout the 1968-79 period, CIDA value biases on trade issues fell clearly into the "vertical constituency portfolio" pattern. Mr. Gerin-Lajoie, CIDA President between 1970 and 1977, described the Agency's role on issues other than aid policy as based on the principle that "CIDA is the single representative which the less favoured countries have in Canadian government councils"[19]. By extension, CIDA's constituency included domestic development-oriented interest groups and churches. Like the Department of Consumer and Corporate Affairs, CIDA's trade positions were straightforward: resist any new protective measures against developing countries and argue for improvement in the benefits to these countries of the General Preferential Tariff. The missionary zeal with which CIDA pressed this viewpoint not surprisingly brought it into frequent disagreement with Finance and IT&C in the interdepartmental committees on the GSP and Low Cost Imports.

Since the arrival of Mr. Michel Dupuy as President in 1977, CIDA's trade positions have become less uncompromising in tone, though the basic attitude of sympathy toward developing country interests has in no way diminished. In contrast to his predecessor, Mr. Dupuy, a career diplomat, is generally believed to be more of a "total government person" than a "CIDA man," and has made an effort to move CIDA into closer alignment with the Department of External Affairs, resulting in more moderate but still liberal CIDA trade positions.

The impact of CIDA on trade decisions during the 1968-79 period was rather limited. This reflects the greater potency of domestic political considerations - and these are particularly acute in situations of import pressure from low-wage developing countries - over internationalist concerns. But CIDA enjoyed some success, in alliance with the Departments of External Affairs and Consumer and Corporate Affairs, in resisting backtracking on the GSP, and moderating the conditions of special import policy measures.

(8) Other Government Departments

In contrast to its American namesake, the Canadian Department of Labour has had a quite modest role in trade decisions, largely because the dispute mediation and labour market policy functions performed by the U.S. department have been split in Canada between two smaller departments, Labour, and Employment and Immigration. Labour Canada was represented on several interdepartmental committees and study groups during the 1968-79 period, including the Low Cost Imports Committee, a panel reviewing the Auto Pact, and the Interdepartmental Committee on Trade and Industrial Policy which studied the need for improved adjustment assistance measures for firms and workers in conjunction with the Tokyo Round. Although Labour falls most closely into the "vertical constituency" portfolio pattern, and is undoubtedly sympathetic to the views of organized labour, it has probably been somewhat less "representational" of labour views in government councils than, say, Agriculture has been in relation to farmers. Thus, while frequently performing the role of facilitator for the channelling of labour views to Cabinet and the Office of the Coordinator for the Multilateral Trade Negotiations, the Department has stopped short of simply mirroring the protectionist claims of many local and regional unions. Because of its work dispute mediation role, it has tried to preserve an element of neutrality between management and unions in those situations, such as when multinationals desire to relocate production abroad, where the interests of the two sides are not complementary. The diverse trade interests of particular unions also militates against straightforward protectionist stands. Reconciliation of these diverse management and labour interests was apparent in Labour Canada's successful efforts to introduce adjustment assistance programs for displaced workers at the conclusion of the Kennedy Round and during the formulation of the 1971 Textile Policy.

The Department and Commission of Employment and Immigration (E&I), known before 1977 as Manpower and Immigration, is responsible for labour market policy, or matching labour supply and demand through training, direct job creation and adjustment assistance programs. Its most consistent involvement in trade issues is the provision of information on the employment impact of trade options on localities and regions. The Canada Employment and Immigration Commission sits on the Low Cost Imports Committee and the Committee on Trade and Industrial Policy, but is seldom consulted about particular tariff issues. E&I has not resisted trade liberalization in situations where relocation and re-employment of displaced workers can be effectively accomplished through government adjustment programs or, preferably, through the private market process. However, in some industries, such as textiles and footwear, where severe cultural and economic bottlenecks hinder smooth adjustment, its positions have been clearly protectionist. This was particularly so during the tenure of Ministers Jean Marchand and Bryce Mackasey, who represented ridings in which these problem industries are located.

The Department of Fisheries and the Environment has played a clear representational role for its constituency. Since Canadian fisheries are little subject to import competition, the Department's trade role has emphasized inputs into the MTN decision-making process in which expanded exports of fish products became a major objective.

The trade role of the Department of Regional Economic Expansion (DREE) stems from its responsibility for the Regional Development Incentives Program, whose grants to such companies as Michelin Tire and to Maritime fishermen have been identified by the United States as subsidization against which countervailing duties have been levied or threatened. DREE also plays a representational role in government for its designated regions, as indicated by its entry into the Committee on Low Cost Imports in 1977 at the request of the Department's Quebec regional office, its studies of the potential impact of MTN tariff reductions on underdeveloped regions of Canada, and its advocacy of a spatial dimension to IT&C industrial sector studies. The Department has hesitated between political pressures encouraging protectionist responses to tariff and quota issues, and the growing realization of the desirability of a regional development rooted in international competitiveness.

(9) The Executive Support Agencies

In the early years of the Trudeau government, several commentators predicted a vastly expanded policy role for the Privy Council Office and the Prime Minister's Office.[20] No doubt the role of these central agencies did increase in the new government's first term of office as Mr. Trudeau's philosophy of counterweights to the influence of the traditional bureaucracy was implemented. As the 1970s unfolded, however, the bureaucracy re-established itself as the primary source of advice to Cabinet on trade issues, based on its clearly superior analytic capability. But PCO and PMO retained special and important roles in the trade decision-making process.

(a) Privy Council Office

PCO's primary function in the machinery of government is to oversee the implementation of the Prime Minister's views on how public decisions should be taken. For this reason, it is often referred to as the "Prime Minister's Department." Mr. Trudeau's approach to decision-making emphasized the principle that all ministers, whether formally responsible for a given issue or not, should have the opportunity to participate in policy-making. Accordingly, PCO coordinates the bureaucratic preparation of memoranda to Cabinet, ensures that appropriate interdepartmental consultation has taken place, and distributes Cabinet papers to all ministers.[21] In addition, the Office is organized to provide secretarial and administrative support to Cabinet Committees, including those which most often consider trade issues, the Committees on Economic Policy, External Affairs and National Defence, and Government Operations.

PCO is not equipped to compete with the Departments of Finance and Industry, Trade & Commerce as a source of substantive analysis on most trade questions. Its policy role and influence have been most visible in the areas of creating organizational frameworks for action and in communicating to the bureaucracy what it perceives as the collective will of ministers. An example of the framework function was provided by the PCO initiative in 1977 to create a new Office of Coordinator for the Multilateral Trade Negotiations to bind together existing interdepartmental, federal-provincial and government-private sector consultative mechanisms. The policy communication role, most often played out by participation of PCO officials in bureaucratic level committees, is a frequent avenue of PCO influence on policy outcomes. In 1976, for instance, PCO effectively scuttled Canadian support for the so-called STABEX plan for stabilization of developing countries' commodity earnings by pointing out that the government had already rejected a similar domestic plan proposed by Canadian farmers.[22]

(b) Prime Minister's Office

The policy role of PMO, the principal source of partisan political advice to the Prime Minister, increased greatly during the Trudeau years, but this trend was less apparent in trade policy than in fields like federal-provincial relations. Probably the most important reason is that Mr. Ivan Head, Trudeau's Senior Advisor for International Relations until early 1978, did not have the manpower to perform on a day-to-day basis a real countervailing role vis-à-vis the bureaucracy. Head had no assistants, and was not particularly expert on trade matters himself.

Head's influence, sporadic but sometimes important, was most visible on two issues generally believed to be of great concern to the Prime Minister: international development and the promotion of trade links beyond the North American continent. He played a conspicuous role in 1975, for example, in persuading the reluctant IT&C and Finance to go along with External's desire for an Interdepartmental Committee on Economic Relations with Developing Countries to coordinate Canada's response to the "new international economic order." Head was instrumental in shifting the Canadian position during UNCTAD IV to an attitude of greater sympathy toward the UNCTAD "integrated commodities program." He was also, with former Secretary of State for External Affairs Mitchell Sharp, a prime mover behind the "Third Option" and the "contractual link" with Europe. On the other hand, Head seldom intervened in specific day-to-day decisions on tariffs, quantitative restrictions and other import barriers, largely because it proved impossible for one man to cover the vast array of foreign policy issues in any detail.

(10) Economic Council of Canada

The Economic Council of Canada was set up by the Pearson government as an independent advisory body reporting to the Prime Minister. Council members are broadly representative of the business, labour and academic communities, while research staff are recruited primarily from academia and the bureaucracy.[23]

The major contribution of the Economic Council of Canada to the debate on trade policy during the Trudeau years was the publication in 1975 of a concensus report, entitled Looking Outward: A New Trade Strategy for Canada[24], and several associated studies. Among other things, the report recommended broad unilateral tariff reductions by Canada to encourage rationalization of Canadian manufacturing, the negotiation of a free trade area with the United States, a vastly increased program of adjustment assistance to Canadian industry, and the amalgamation of the Tariff Board, Anti-Dumping Tribunal and Textile and Clothing Board into a larger trade review agency along the lines of the United States International Trade Commission.

The bureaucratic and political reaction to Looking Outward was quite harsh. The proposals for unilateral tariff cuts and free trade with the United States proved particular sore points. One interviewee, a former Cabinet minister, replied that while the report was discussed in Cabinet, there was not a single minister with any sympathy for it because "none would want to fight an election on the issue of free trade with the United States."[25] While it may be true that some politicians are privately sympathetic to the idea of a Canada-United States free trade agreement (the 1977 Canadian Elite Image Study having found about a third of ministers interviewed to be in favour)[26], public proclamations in favour of such ideas are generally limited to ex-politicians, such as Robert Stanfield.[27] Reaction in the bureaucracy to Looking Outward was equally negative. The general feeling, especially in Finance and IT&C, was that the report was logically impeccable but politically naive. Finance officials were especially upset by the Council's advocacy of unilateral tariff cuts at a time when Canada was negotiating internationally.

Economic Council reports, of course, tend to be directed at medium to long-term solutions to problems. This is at once their major contribution to the policy process and their greatest weakness. Given that they tend not to focus on short-term policy options, they run the risk of either being ignored by politicians and bureaucrats, or dismissed as impractical. On the other hand, reports such as Looking Outward can be useful stimulants to public and intra-governmental debates on important problems.

NOTES: Chapter Five

1. G. Bruce Doern, "Horizontal and Vertical Portfolios in Government," in G.B. Doern and V.S. Wilson (eds.), Issues in Canadian Public Policy (Toronto: McGraw-Hill Ryerson, 1974) Chapter 12.

2. Ibid., p. 316.

3. Ibid., p. 318.

4. Peyton V. Lyon and David Leyton-Brown, "Image and Policy Preference: Canadian Elite Views on Relations with the United States," International Journal, Vol. 32, No. 3 (Summer 1977) pp. 640-671.

5. Richard W. Phidd and G. Bruce Doern, The Politics and Management of Canadian Economic Policy (Toronto: Macmillan, 1978) p. 150.

6. A complete and recent listing of these programs is available in Board of Economic Development Ministers, Assistance to Business in Canada (Ottawa: Minister of Supply and Services Canada, 1979).

7. Mr. Pitfield's philosophy of government is articulated in his article, "The Shape of Government in the 1980s: Techniques and Instruments for Policy Formulation at the Federal Level," Canadian Public Administration, Vol. 19, No. 1 (Spring 1976) pp. 8-20.

8. The 23 reports were prepared by task forces composed of representatives of the industries, a few academics, and observers from the federal and provincial governments. The overall conclusions of the exercise may be found in A Report by the Second Tier Committee on Policies to Improve Canadian Competitiveness (Ottawa: Department of Industry, Trade and Commerce, October 1978).

9. The five Ministers of IT&C in the Trudeau period were: Jean-Luc Pépin: July 1968 to November 1972; Alistair Gillespie: November 1972 to September 1975; Don Jamieson: September 1975 to Sept-

ember 1976; Jean Chrétien: September 1976 to September 1977; and Jack Horner: September 1977 to May 1979.

10. This preoccupation was justified. Mr. Horner was defeated in the general election of 22 May 1979.

11. Economic Council of Canada, Looking Outward: A New Trade Strategy for Canada (Ottawa: Information Canada, 1975).

12. Confidential interviews.

13. Bruce Thordarson, Trudeau and Foreign Policy (Toronto: Oxford University Press, 1972).

14. See Canada, Department of External Affairs, Foreign Policy for Canadians (Ottawa: Information Canada, 1970).

15. W.M. Dobell, "Interdepartmental Management in External Affairs," Canadian Public Administration, Vol. 21, No. 1 (Spring 1978) p. 90.

16. The "third option" was the middle ground proposed by a 1972 government study on Canada-U.S. relations, between the "extremes" of the status quo (the "first option") and further continental economic integration (the "second option"). It aspired to progessively diversify Canada's international relationships, particularly by expanding trade with Europe and the Pacific Basin states. See Mitchell Sharp, "Canada-U.S. Relations: Options for the Future," International Perspectives, Special Issue, Autumn 1972.

17. See Tariff Board, Report on Fresh and Processed Fruit and Vegetables, Volumes 1-5, Ref. No. 152 (Ottawa: Minister of Supply and Services Canada, 1977-78).

18. For Mrs. Plumptre's conclusions, see Food Prices Review Board, Final Report: Telling it like it is (Ottawa: February 1976).

19. Paul Gerin-Lajoie, "CIDA in a Changing Government Organization," Canadian Public Administration, Vol. 15, No. 1 (Spring 1972) p. 54.

20. See G. Bruce Doern, "The Development of Policy Organizations in the Executive Arena" in G. Bruce Doern and Peter Aucoin (eds.) The Structure of Policy Making in Canada (Toronto: Macmillan, 1971), and Thomas D'Aquino, "The Prime Minister's Office: Catalyst or Cabal?", Canadian Public Administration, Vol. 17, No. 1 (Spring 1974) pp. 55-79.

21. An excellent account of the work of the Privy Council Office is provided in Richard D. French, "The Privy Council Office: Support for Cabinet Decision Making" in Richard Schultz, Orest M. Kruhlak and John C. Terry, (eds.) The Canadian Political Process, 3rd ed. (Toronto: Holt, Rinehart and Winston, 1979) pp. 363-394.

22. The STABEX scheme is an agreement between the European Economic Community and its associated developing country members to stabilize prices of the latter countries' commodity exports to Europe.

23. For a broader discussion of the Council's role in making Canadian economic policy, see Richard W. Phidd, "The Economic Council of Canada, 1963-1974," Canadian Public Administration, Vol. 18, No. 3 (Fall 1975) pp. 428-473.

24. Economic Council of Canada, Looking Outward, op. cit.

25. Confidential interview.

26. Peyton V. Lyon and David Leyton-Brown, op. cit.

27. Robert Stanfield, "Exploring a New Common Market," Globe and Mail, 8 February 1978.

CHAPTER SIX

CASE STUDY: QUANTITATIVE RESTRICTIONS AND LOW COST IMPORTS

Governments of industrialized countries, since the Kennedy Round's substantial success in lowering tariffs, have increasingly resorted to quantitative restrictions on imports. This chapter presents a case study of the Canadian decision-making process for "special import policy" as it relates to quantitative controls on "low cost" manufactured goods, particularly textiles and clothing.[1]

In terms of the decision models described in Chapter One, it will be argued that in the majority of the 50-odd cases of textile restraints imposed by Canada between the announcement of a new Textile Policy in 1971 and the end of 1976, decisions were arrived at in close conformity to the "governmental politics" model. In a smaller number of cases, however, this standard or typical dynamic was replaced by a "rational-political" process. A major example of the rational-political, or simply "political" model, took place in November 1976, when the Minister of Industry, Trade and Commerce announced global quotas on the importation of 14 items of clothing affecting some $600 million in trade.[2] Since this decision amounted to the most serious imposition of a non-tariff barrier to trade by Canada since the Depression, we shall examine it here in some detail.

(1) Background to the Textile Import Problem

As the economics of the troubled Canadian textile industry have been fully documented elsewhere,[3] we need only mention a few of its most salient characteristics. This industry can be divided into three major components: a chemical sector (production of man-made fibres); a primary sector (weaving, spinning, knitting of various fabrics); and an apparel or clothing sector (cutting and sewing of fabrics into garments). The linkages between the sectors are such that some 45 percent of the primary industry's production is used as input into the apparel sector, and thus the primary firms share with apparel firms an interest in import restrictions on finished garments. This dependency relationship does not operate in the reverse direction: apparel firms do not share the primary sector's interest in protection against fabric imports, since they wish to obtain inputs at lowest cost, regardless of source.

Both primary and apparel sectors are concentrated in Quebec (69.7 percent of apparel firms) and Ontario (21.3 percent), with smaller pockets of activity in Manitoba, British Columbia and Nova Scotia.[4] The textile industry is very labour-intensive, employing directly some 114,000 persons in 1976 in the clothing sector and 171,000 in the primary sector. These 285,000 jobs represent, in terms of employment, the second largest manufacturing sector in Canada. In Quebec, one in four manufacturing jobs is in the textile industry. Compared with other manufacturing sectors, the textile industry has a relatively immobile labour force; some 75 percent are female, and 40 percent of the force in Quebec are immigrants with few alternative job prospects. Compared with other countries, labour costs in Canada are very high. They are, in fact, surpassed only by Sweden:

Hourly Wages: Clothing Industries, August 1976 ($U.S.)

Sweden	$6.82	Japan	$1.60
Canada	4.40	Poland	0.88
U.S.	4.05	Romania	0.67
France	3.12	Taiwan	0.48
U.K.	2.22	S. Korea	0.38

Source: Textile and Clothing Board, Clothing Inquiry, Ottawa, May 29, 1977, p. VI-9).

It is against this background that the "import problem" should be viewed. During the 1950s and 1960s, a number of developing countries, primarily in East Asia, as well as some East European state trading countries, developed the capacity to export quality fabrics and garments. Although often subsidized by government incentive programs, the key to their competitive success was of course comparatively low wages. This phenomenon caused severe dislocations in the domestic textile industries of the industrialized countries. The latter reacted to these pressures in a two-pronged way: by negotiating multilateral trade agreements over and above the GATT to govern the international exchange of textiles; and by creating a series of "national textile policies."

Tariff rates and other protective devices such as customs valuation procedures no longer provided adequate protection to domestic producers faced with this surge of Third World and East European imports. Canada's nominal tariff on textile products averages 17 percent and on clothing 21 percent, while effective rates reach 24 and 25 percent respectively.[5] This amounts to one of the highest degrees of tariff protection afforded any Canadian industry. Yet the Third World demonstrated a rising ability to jump this tariff wall, as indicated by these figures on the share of the Canadian clothing market held by Canadian manufacturers:

Percent of Canadian Market (in volume)

	1971	1973	1974	1975	1976
Domestic	73%	67	68	64	55
Imports	27%	33	32	36	45

Source: TCB, Clothing Industry (Ottawa, May 29, 1977, p. II-3).

GATT provisions proved inadequate to deal with surges in textile exports from low cost countries. The GATT could not provide for an orderly expansion of Third World exports while at the same time allowing Western domestic industry the time and security to plan adjustment to this competition through specialization and rationalization. The GATT "safeguard" clause, Article 19, permits only non-discriminatory restrictions, regardless of the source of injury, which on the whole stemmed from low wage countries. This led to a proliferation of restrictions in the 1950s which hurt other industrialized countries that were not normally the source of injury, and to other "illegal" restrictions which discriminated against developing countries. Exporting countries were also adversely affected in that restrictions once in place tended to remain indefinitely, and were imposed in an arbitrary manner without consultation or review. In order to head off a rapidly gathering protectionist movement in many developed countries, the Long Term Arrangement Regarding International Trade in Cotton Textiles (LTA) was negotiated in 1962, lasting until 1973. The LTA was a departure from the GATT non-discrimination or MFN principle in that importing countries were permitted to introduce restraints on cotton imports against specific countries, generally via negotiated "voluntary export restraints agreements." During the 1960s, however, fibres other than cotton came to dominate the international textiles trade and many countries, including Canada, reverted to Article 19 GATT action with all its defects. The result was the Arrangement Regarding International Trade in Textiles, also known as the Multifibre Agreement (MFA), which came into effect in January 1974. The MFA covered all textile fibres and again provided for discriminatory restrictions against the specific countries deemed responsible for injury. It also added several improvements from the standpoint of Third World producers. Restrictions were to be limited to specific product classes causing or threatening injury, for specified periods of time, in consultation with exporting countries, and only after public inquiries had documented the presence of injury from imports. An international mechanism, the Textile Surveillance Body, was created to which restrictions were to be reported. The MFA also provided for more favourable base levels and growth rates for developing countries[6], and exempted handloom and handicraft textiles from restraints.[7]

Besides multilateral accords, Western governments also reacted to import pressures by adopting industrial strategies for their domestic textiles sectors. Sweden and Japan went furthest in phasing out parts of their domestic operations, while the EEC and the United States pursued unabashedly "closed door" approaches to the problem. The United States, for example, adopted a policy of comprehensive long-term import controls, normally for five years, with the result that domestic producers maintained an 80 percent share of the American market in 1976. The Canadian government's policy, generally referred to as the "Textile Policy," was introduced by former IT&C Minister Jean-Luc Pépin in 1970-71.[8] The policy had the following characteristics:

(i) It was based on the assumption that the Canadian industry could become technologically efficient enough to meet import competition from other industrial countries with the aid of the tariff alone, and indeed would be able to improve its export performance. In contrast to previous Canadian approaches to the import problem, the new policy attempted to balance producer and consumer interests by highlighting the concept of "rationalization" of production processes, suggesting that special protection above and beyond the tariff would be conditional upon improvements in companies' rationalization performance. With this in mind, a number of industrial assistance and export promotion programs were set in motion. While this appeared a realistic approach to meeting competition from other developed countries, the policy was never clear as to how much contraction of the Canadian industry would be permitted to accommodate the labour cost advantage of developing countries, or to what degree it would accept wage differentials as a legitimate element of comparative advantage that should not be interfered with by commercial policy measures. In times of unemployment and political pressure from the industry, the tendency has been to see "cheap labour" advantages as inherently unfair and to set aside the strict interpretation of conditionality.

(ii) Special measures of protection would be used to prevent injurious imports from low wage countries during the period of rationalization. Specifically, the Export and Import Permits Act was amended to empower Cabinet to use quotas and voluntary export restraints by placing goods "causing or threatening to cause serious injury" on the Import Control List.

(iii) The government set up formal mechanisms to implement the policy. The Textile and Clothing Board (TCB) was created to conduct inquiries as to whether injury or threat thereof was occurring, and was empowered to recommend special measures of protection only if it was satisfied that the firms involved had undertaken plans for adjustment to meet international competition. The chairmanship of the Interdepartmental Committee on Low Cost Imports was transferred from the Department of Finance to IT&C. The Committee was charged with interdepartmental review of TCB recommendations and with making recommendations to the Minister of IT&C. Ultimately, special import policy was decided in Cabinet or Cabinet Committee. The Office of Special Import Policy of IT&C was charged with negotiation of restraint agreements with exporting countries.

(iv) At least until November 1976, Canadian practice on import restraints was largely in accord with the Multifibre Agreement, in contrast to some other importing countries. Voluntary export restraints were negotiated only after public inquiry to establish whether in fact injury or threat of injury had arisen. Controls were imposed only against those countries responsible for the injury, for specific products shown to have caused injury, and for specified periods of time, with the restraints usually lifted after the threat subsided.

The Pépin textile policy resulted in considerable technical rationalization of the Canadian industry, but it could not arrest the trend toward ever increasing import shares of the Canadian market. Technical rationalization simply could not make up for the labour cost disadvantage of Canadian producers. Among Western countries, only Sweden had experienced deeper import penetration of its textile and clothing markets. From a domestic political perspective, the Canadian government's attempt to establish a delicate balance between reasonable protection for a troubled Canadian industry and providing the consumer with a wide range of reasonably-priced textiles and clothing pleased neither manufacturers, nor importers and retailers, nor consumer groups. Furthermore, many academic economists questioned whether the Textile Policy was realistic in assuming that time to accomplish technical rationalization could overcome the Canadian industry's wage rate disadvantage and suggested that most of the industry was in fact a lost cause.

(2) The Decision-Making Environment: Pressures and Constraints

In this section we shall take a brief look at the views and behaviour of domestic interest groups, provincial and foreign governments and Members of Parliament on the textile import question in the 1970s. These pressures and constraints appear to fit rather neatly into Walter's macro-model of trade policy influences (described in Chapter One), with actors being divided into "protection-biased" and "trade-biased" sectors.

The Canadian textile industry had long argued that it was as technologically efficient as any in the world, but that uncertainty about future levels of imports from low-wage countries severely constrained its ability to plan future investments. Responsibility for this uncertainty was placed squarely on the shoulders of the federal government, which was accused of long delays in the negotiation of restraint agreements and too great a "wish to adhere scrupulously to international rules."[9]

Because of the speed with which they can be implemented, the textile industry as a whole repeatedly urged greater use of GATT Article 19 global quotas on a reluctant federal government that generally preferred the more selective approach of using voluntary export restraint agreements to minimize upsetting Canada's major trade partners. But the primary and apparel sectors parted company on the product by product application of these quotas. The Canadian Textiles Institute, the major representative of primary producers, recommended a comprehensive approach to protection - three tier quotas to cover fibre, fabric and finished garment imports. The clothing manufacturers on the other hand, led by the Apparel Manufacturers Association of Quebec, and supported by the

Ontario, Manitoba and B.C. associations, desired long term quotas on garment imports, but wished to limit restrictions on fabrics used as inputs into their production processes.

Predictably, the major textile trade unions - the Centrale des Syndicats Démocratiques, the Confederation of National Trade Unions, the Amalgamated Clothing and Textile Workers Union and the International Ladies' Garment Workers Union - strongly supported their employers' view of the unfairness of competition from "cheap labour" Third World and state trading countries. In fact, the textile unions had for several years been at odds with the generally pro-free trade stance of the Canadian Labour Congress, and had unsuccessfully pressured the CLC to change its policy on international trade.

Strongly opposed to these views were the importers', retailers' and consumers' associations. About the only issue on which they had traditionally agreed with manufacturers was in condemning government for its "total lack of direction."[10] The Canadian Textile Importers Association, the Canadian Importers Association and the Retail Council of Canada (representing the major department stores) all agreed that import controls would involve a major cost to the consumer. They further argued that the efficiency of the Canadian textile industry was in serious doubt and that at least part of it should be phased out. Importers and retailers agreed with the manufacturers that, if quotas must be imposed, they should be long term and global (to allow maximum "shopping around"). But unlike the manufacturers, they called for as large quotas as possible and that they should be made strictly conditional on improvements in the efficiency of Canadian industry. The general thrust of the importers' and retailers' arguments was accepted by the Consumers' Association of Canada (CAC), which was particularly concerned about the price and availability of clothing for low income consumers.

Provincial governments also intervened on particular special import policy issues, most frequently to defend local employment interests. The governments of

Quebec, Ontario and Manitoba frequently made statements critical of federal policy as overly generous to foreigners. Their representatives on the IT&C Minister's Advisory Committee on Textiles gave consistent support to industry viewpoints.[11] While the general views of non-producing provinces were well known to favour a liberal import policy, they seldom intervened on particular issues or on Textile and Clothing Board reports, leaving this role to their regional representatives in the federal Cabinet.

Foreign governments, through bilateral contacts and in the GATT Committee on Textiles, intermittently commented on Canadian policy, criticizing protective actions. There is no doubt that these views represented a major constraint upon the severity of Canadian textile import policy, particularly on the use of GATT Article 19 global action; global restraints would affect the interests of major trade partners like the United States which were in a position, unlike most of the developing country exporters, to retaliate against Canadian protectionism.

Finally, parliamentary actors were an important influence on many textile issues, particularly individual MPs reacting to pressures from riding interests, and exerting a "protection-biased" impact, either in the governing party caucus, in the Standing Committee on Finance, Trade and Economic Affairs, or from the opposition benches. This intervention cut across party lines, most of the affected MPs being Liberals in Quebec and Conservatives in Ontario. Informed observers assess the influence of individual MPs in favour of protective action, particularly government caucus members, as frequently considerable.[12]

Other parliamentary institutions periodically commented on these issues. One was the Subcommittee of the Standing Committee on External Affairs on International Development, whose Report to the House of April 1976 advocated liberalized trade policies toward developing countries[13]; another was the Senate, whose hearings on the textile industry and report (April 1976) proved very supportive of the domestic industry's view.[14] The Senate hearings, as we shall see, did have some impact on the November 1976 decision to impose global quotas on clothing, while the influence of the House Subcommittee can best be judged by the comment of a senior member of the Textile and Clothing Board that he had not heard of its report.[15]

(3) The "Standard" Special Import Policy Process

Between the creation of the Textile and Clothing Board in May 1971 and the end of 1976, some 50 low cost import cases passed through a fairly formal decision-making process set up to consider these issues as they arose. The process involved three major stages: review of interest group claims and preparation of a report by the Textile and Clothing Board; consideration of the Board's report and preparation of recommendations to Cabinet by an interdepartmental committee; and final decision by Cabinet Committee.

There existed a "typical" or standard decision-making dynamic to this process, best described by the "governmental politics" model, and which was in clear contrast to the manner in which the November 1976 decision on clothing imports was handled. Our research was not conducted in such detail that we are able to say that the governmental politics model characterized all decisions before November 1976, but interviews indicated that this "typical" dynamic worked in the great bulk of cases between 1971 and 1976.

(a) Stage 1: Textile and Clothing Board

In its reports on industry claims of "injury or threat of injury," the Board, with great regularity, has found injury to be occurring, and appealing firms' rationalization plans to be acceptable. Accordingly, it has usually recommended the introduction of special measures of protection. As was argued earlier in the study, the Section 18 principle of the Textile and Clothing Board Act - that special import restraints should not be used in the long term to maintain lines of production which require more than customs duties to remain price competitive - appears to have been largely set aside in the Board's reports as import pressures from low-wage countries mounted.

(b) Stage 2: Interdepartmental Review

Created by order in council in 1961, the chairmanship of the Interdepartmental Committee on Low Cost Imports (LCI Committee) passed from Finance to IT&C in 1970 as part of the Textile Policy. According to the Cabinet directive authorizing this transfer, the Committee's mandate is essentially to review Textile and Clothing Board reports and to recommend to Cabinet whether special import controls are necessary, and the form they should take.[16] Legally speaking, these recommendations amount to determining just what category of Import Control List measure (i.e., Article 19 global action, negotiated export restraint agreements, or import surveillance without actual controls) should be invoked, and how the provisions of the Multifibre Agreement on base levels and growth rates should be distributed among exporting countries. The LCI Committee was also empowered to recommend import surtaxes for problems deemed to be of a more short-term nature.

The Committee was specifically directed by Cabinet to consider a variety of conflicting concerns, including international commitments under the GATT and Multifibre Agreement and diverse domestic consumer and employment interests. In undertaking the difficult assignment of reconciling these objectives, however, a major weakness of the LCI Committee mechanism has been that it does not have access to company plans for adjustment, the main condition for protection under the Textile Policy. The *Textile and Clothing Board Act* ensures the confidentiality of company plans submitted to the Board. Although some officials in other departments suspect that the higher IT&C officials have access to these plans, there is no hard evidence to substantiate such claims. It is quite possible that IT&C officials are able to obtain a rough idea of the plans through contacts between companies and the Textile and Consumer Products Branch of IT&C, but they are not getting the information from the Board. One senior IT&C official admitted he had requested it at least once, but insisted that it was refused.[17] Many officials in all departments, including IT&C, have their doubts about the seriousness of company adjustment plans. These suspicions are impossible to verify, but the fact that officials voice them affects the nature of the decision-making process on low cost imports. Incomplete data and less than full confidence in the Board encourages an interest representation process in the Committee, particularly on the part of "vertical constituency" departments.

Another problem faced by the LCI Committee has been difficulty in objectively defining "serious injury or threat of injury," the other major condition for protective action under the Multifibre Agreement. It is hard to determine at a specific time whether the problems of the industry are the result of unfair imports, or whether factors such as a down-turn in the economy, a shift in consumer demand, or poor management are predominant. The Canadian industry, in any case, is always grumbling about imports. The absence of complete information on plans and objective criteria for establishing injury reduces the scope for decision-making based on informed and considered analysis. Instead most member departments and sub-units of departments have tended to put forward subjective arguments based on the domestic or international interests they see themselves as representing. Decisions thus seem to be based as much on the dynamics of bargaining between departments as on informed analysis.

The LCI Committee has as members the departments of IT&C, Finance, External Affairs, Consumer and Corporate Affairs, National Revenue, Employment, DREE, Labour, the Privy Council Office, and the Canadian International Development Agency. As in most interdepartmental committees on trade questions, the ministries of IT&C, Finance and External Affairs stand out as far and away the most significant, to the point that"if these three agree, the measure proposed is almost certain to be recommended to the Minister."[18] No formal votes are taken, but the other departments are able to influence outcomes in the event of division among the heavyweight departments, which is quite frequent. The Committee functions officially at the assistant deputy minister (ADM) level, but like most other such mechanisms, much work is done at a lower "working level."

Decision-making in the committee normally begins with a set of IT&C proposals based on the recommendations of the Textile and Clothing Board. The typical interaction is then "for Finance and External to attempt to whittle down IT&C's protectionism."[19] The others, in particular Consumer Affairs and CIDA, attempt to do the same, but generally in a way perceived by the big three as more one-sidedly in favour of their special constituencies. The LCI Committee is not a simple rubber stamp for either IT&C or the Textile and Clothing Board. IT&C officials, especially those from the Office of General Relations, have themselves proven reasonably open to change, but this flexibility depends greatly on the attitude of the incumbent Minister and the extent and form of pressure he puts on his own officials. Alistair Gillespie, for example, was generally considered less firm in his opinions on textile questions than was Jean Chrétien. Thus there was more room during Gillespie's tenure for interdepartmental compromise. The interest and determination of other ministers in Cabinet is of course also relevant.

The other side to the decision-making dynamic of the Committee, and one which imposes upper limits on the amount of interdepartmental bargaining that takes place, is the lead department principle. Where one department has clear responsibility in a given policy area there exists a sort of gentlemen's agreement accepting its legitimacy or primacy. Other departments, but especially the big three, will not press their case too far, thus avoiding intractable disputes.

The Minister of IT&C is not bound to accept the recommendations of the LCI Committee and has on several occasions rejected them. For example, a Committee recommendation in 1975 to impose a surtax on polyester yarns was rejected by Alistair Gillespie, only to be imposed later by Jean Chrétien. However, most recommendations agreed to by IT&C, Finance and External have been accepted.

Departmental concerns and behaviour in the LCI Committee correspond roughly to the characteristics profiled earlier. By the time issues get to the Committee, some thrashing out of conflicts has usually taken place within IT&C between the differing orientations of the Textile and Consumer Products Division and the International Trade Relations Branch. Within the ITR Branch, the Office of General Relations (OGR) is generally more wary of protectionism than the Office of Special Import Policy, whose function it is to negotiate restraint agreements with exporting countries. The results of this internal IT&C decision-making process have generally been to call for some increased special protection in the draft memorandum to Cabinet, reflecting the influence of the Textile and Consumer Products Division, but in a more moderate form than the original Textile and Clothing Board recommendation, in line with the preferences of OGR.

The Department of Finance, whose International Economic Relations Division serves on the committee, does not represent any one interest on low cost import questions and its positions reflect a conscious attempt to balance the diverse employment, inflation and diplomatic dimensions to these issues. Consequently, its positions normally fall between those of other departments, generally accepting the need for protective action, but in the least blatant and most conditional form possible. Finance's prestige and recognized expertise, as well as its inclination to play the role of reconcilor of competing concerns, have contributed to its considerable success in liberalizing IT&C's original proposals, particularly in exacting concessions on growth rates, type and duration of protection and rationalization conditions.

External Affairs views its role in the LCI Committee as essentially one of promoting consistency of foreign policy objectives. Its concern for the Canadian image on international development questions, and sensitivity to the views of foreign governments and international obligations, causes its positions to be quite liberal. On the other hand, External has proved noticibly more sensitive to domestic employment and political considerations than the liberal "vertical constituency" departments on the Committee, Consumer and Corporate Affairs and CIDA. Generally speaking, External's stances have been quite similar to those of Finance, but perhaps somewhat more liberal.

At the same time, External's disinclination to upset other governments sometimes conflicts with its general sympathy toward developing countries. The Multifibre Agreement permits importing countries to give special treatment to lower income developing countries on base levels and growth rates of quotas and negotiated agreements. CIDA regularly proposes such action. More intermittently, Finance and the ITR Branch of IT&C agree to special treatment as a compromise to conflicts within the committee and because most of the lower income developing countries represent less of an import threat than countries like Hong Kong and Korea. But External opposes special treatment out of a desire not to be seen playing favourites.

The views of other committee members are consistent with their "vertical constituency" roles. Although it has not been particularly active, the Department of Labour, like National Revenue and Employment and Immigration, can be counted upon to support protection. Consumer and Corporate Affairs (CCA) and CIDA are the Committee's most consistently and intensely liberal members. Their influence was probably diminished for several years by a tendency to adopt extreme, "all or nothing" tactics in Committee deliberations, an approach disparagingly referred to in IT&C as the "what-can-we-give-away-today" policy.[20] In contrast to Finance and External, which have accepted the premises that IT&C's basic proposals cannot be altered, and that a limited strategy of obtaining concessions should be their aim, CCA and CIDA often argued against any protection at all, a stance regarded by other participants as unrealistic. The reason for CCA's and CIDA's intemperance was probably that these two departments were really the only ones possessing fairly coherent trade "ideologies," one based on free market competition and the other on a world-view calling for a "new international economic order." More recently, CIDA and CCA appear to have become more pragmatic. CIDA, for example, has since 1976 put priority on achieving concessions for countries which are recipients in the Canadian aid program, and is more willing to accept restrictions against higher income developing countries.

The most common outcome of this "standard" LCI Committee process between 1971 and 1976 was a compromise in which the relatively protectionist departments and units got their way as to whether protection per se would be afforded but where the liberal forces won concessions on the conditions of protective actions (e.g., base levels, growth rates, type and duration of protection, and rationalization conditions). The precise degree to which the liberal and protectionist elements could claim victory was highly influenced by the officials' anticipations of the ministers' reactions to options considered in the Committee. It was already pointed out that the tenure of the relatively neutral Mr. Gillespie was associated with the extraction of greater concessions by the more liberal forces as compared with the period in which Jean Chrêtien was Minister. The anticipated reactions of other ministers was also relevant, however. External Affairs and Consumer and Corporate Affairs ministers, for instance, varied over the period in their interest in the issue of low-cost imports and in their commitment to the various dimensions of the problem. External and CCA officials were firmer in the committee when they felt they had full support from their Ministers. Probably the most salient example was the case of Allan MacEachen as Secretary of State for External Affairs, whose strong commitment to international development was widely recognized.

(c) Stage 3: Cabinet Committee

In the "typical" low cost import process, officials in the interdepartmental committee proved quite successful in their anticipation of ministerial preferences and most memoranda passed through the Cabinet Committee on Economic Policy without serious amendment. Interviews indicated that it was the regional dimension to special import policy issues that received most attention at Cabinet level[21]: such concerns are of course not guaranteed to result from interactions of functionally defined departments at the bureaucratic level.

Kelly's dissertation on the formulation of the Textile Policy and the Textile and Clothing Board Act in 1970-71 provides an example of the kind of bargaining and compromise which takes place on these issues at Cabinet level:

> Interviews with senior IT&C departmental officials suggested that Section 21 of the Bill, providing adjustment assistance for textile workers displaced due to rationalization, was substantially strengthened in order to gain the support of the Minister of Manpower and Immigration. Furthermore, it was generally acknowledged that the Minister of IT&C faced some stiff opposition from his Western Canada colleagues to any aspect of the Bill which suggested that special measures of protection would be given to an (Eastern) Canadian secondary industry which might in any way jeopardize (Western) export trade to the Pacific Rim countries.[22]

The workings of governmental politics at Cabinet level, focussing primarily on regional interests, sometimes led between 1971 and 1976 to further watering down of Textile and Clothing Board recommendations. Cabinet's preference in this period for negotiated restraint agreements, as compared with the less selective GATT Article 19 instrument, was also indicative of regional and other compromises in Cabinet.

Meanwhile, imports continued to rise and the textile industries complained about the length of time required to accomplish these compromises within Industry, Trade and Commerce, in the LCI Committee and in Cabinet. Matters were to boil over in the summer and fall of 1976.

(4) The Clothing Decision of November 1976

In 1976 there was a rapid growth in import penetration of the clothing market from low-cost countries. According to the Textile and Clothing Board in May 1977, "in 1976, imports increased at such a sharp rate and in such a disorderly manner that the market was completely disrupted. Among the garments included in this inquiry, the overall increase for 1976 was 46 percent, while the increase among the individual categories ranged from 9 percent to 370 percent."[23]

By the fall of 1976, almost all concerned parties agreed that something had to be done. Yet one must go beyond the fact of rising import penetration to explain the drastic and unprecedented action of November 29th, 1976. It was drastic in several ways. Although not the first time GATT Article 19 quotas had been used by Canada, this global action entailed at least a degree of contravention of the country-specific principle of the Multifibre Agreement. The product-specific rule of the Agreement was also set aside, for the November quotas applied to all 14 clothing items considered in the Textile Board's report, regardless of whether import penetration was closer to the 9 percent lower range or in the upper range. In addition, several existing voluntary export restraint agreements, and another in the process of negotiation, were terminated, and the quotas were imposed without prior consultation with exporting countries or the GATT Secretariat. Although none of these actions were strictly illegal under the Multifibre Agreement, they did represent a clear departure from Canada's usual close adherence to the spirit of the accord.

Our account of the "standard" decision making process on low-cost imports would lead one to expect a rather different approach to the problem of rising clothing imports in 1976. The interdepartmental committee would no doubt have reached various compromises on the country and product application of restraints. Some Cabinet Ministers would have strongly opposed on grounds of regional interests such massive doses of protection for an industry primarily located in Quebec. External Affairs would in all likelihood have insisted upon the diplomatic niceties of consultation with exporting countries, and the whole process would probably have taken several months to resolve.

Instead, the summer and fall of 1976 witnessed an unprecedented politicization of the textile and clothing issue. Ministers reacted to what came to be perceived as a political crisis by setting aside both the usual strong decision-making role of departmental bureaucrats and their own concerns as heads of departments and regional representatives in Cabinet. Their roles as party politicians came to define their approach to the problem. The most reasonable explanation of the outcome is therefore derived from the rational-political (or "political") model of decision-making.

The chronology of this politicization was as follows:

(i) Politicization began March 1976, during Senate hearings on the textile industry, and was given a boost by the Senate Standing Committee on Banking, Trade and Commerce Report of April 7th. This report was highly critical of government policy, claiming that "the cause of this situation appears to be an unwillingness by the government to implement the textile policy in accordance with the aims of that policy, and in fact to promote almost an open door policy."[24] Perhaps of greater significance than the Report itself were the hearings preceeding it. They provided the industry with its first opportunity in some time to "sound off" to government. The new Minister of IT&C, Don Jamieson, who at that point was not especially familiar with the industry, came under heavy fire at the hearings and was determined to find out what was going on. He instructed his officials to institute an IT&C Ad Hoc Committee to study the situation with textile and clothing manufacturers, labour unions and provincial representatives. Neither importers nor consumer groups were asked to participate.

(ii) In June, the official Opposition, led by MP's Heward Grafftey and Flora Macdonald and supported by the Créditistes, introduced in the House of Commons a motion condemning government textile policy. Jamieson's response that "things are not as bad in the whole industry as has been said" irked the producers' associations.[25] Later that month, the Ontario Minister of Industry, Claude Bennett, entered the fray with a denunciation of Ottawa's policies, claiming that some 4200 Ontario jobs were in immediate danger.[26]

(iii) In July an instance of what was widely perceived as government mismanagement came to light; it concerned restraints on imports of outerwear. The fact that the decision came some four months after the Textile and Clothing Board recommendations were issued was cited time and time again during the summer and fall by industry spokesmen to illustrate government insensitivity to their problems and the slowness of the usual decision-making process.

(iv) On July 30th, the Ad Hoc Committee presented its report to Jamieson. It recommended an urgent TCB inquiry into the clothing import situation and the setting up of a permanent Advisory Panel of government, industry and labour representatives. The panel's first task would be to study the desirability of the "comprehensive" approach to import control (i.e., long term bilateral agreements on related products). Jamieson had already rejected the recommendation of industry and labour members of the Ad Hoc Committee that the comprehensive approach should be immediately negotiated. Instead, he called for further study via the Advisory Panel. He also dissociated himself from their statement that import penetration beyond 1975 levels "cannot be tolerated."[27] Consequently, a feeling grew among industry leaders that the government was procrastinating. But Jamieson did promise that the Textile and Clothing Board would soon be requested to inquire into the clothing situation as soon as government-industry consultations had determined what precise items of clothing should be investigated.

(v) A key event in the unfolding drama was the Cabinet shuffle of September 14th. It included the appointments of Jean Chrétien as IT&C Minister and Don Jamieson as Secretary of State for External Affairs. Just two days later, the Textile and Clothing Board received a formal request from Chrétien for an urgent inquiry into the clothing situation, with an interim report as soon as possible.

In view of Mr. Chrétien's prominent role in the November 29th decision, some examination of his views and personal style is useful. As a Quebec Minister with substantial textile interests in his riding, Chrétien was more familiar with the problems of the industry than most ministers, and acutely aware of its political sensitivity. No doubt the defeat of his Quebec predecessor in the IT&C portfolio in the 1972 general election contributed to this awareness. Mr. Pépin's defeat was widely interpreted as in part attributable to local opposition to the Textile Policy. In a speech to the Canadian Textiles Institute in Montreal in early November, Chrétien described as his "priority number one" the safe-guarding of textile jobs and stated that this would involve a cost, of which consumers "will have to take the brunt."[28] He was careful, however, to note that Canada should do its share for developing countries and warned the industry that "I wouldn't want you to try to stick it to the consumer."[29] These latter assertions may have been somewhat perfunctory in that Chrétien had on several occasions described Canada as "the boy scout of the world" with reference to the international development and obligations aspects of textile questions.[30]

Mr Chrétien's personal style may also be relevant to explaining the November clothing decision. Often impressing journalists and others as a man of action fighting the tentacles of bureaucracy, Chrétien informed reporters upon assuming the IT&C portfolio that "I intend to cut out the bureacratic squabbles that often make the decisions six months late."[31] His Montreal speech indicated that Chrétien was about to "develop policies in the next few weeks to give you [i.e., the industry] stability."[32]

(vi) Meanwhile, during September and October, the Textile and Clothing Board (TCB) was conducting its inquiry and preparing an interim report, via meetings with industry and government policy-makers. The most significant part of the interim TCB inquiry, and part of the increasing politicization of the issue during the summer and fall of 1976, was a joint representation by the clothing associations under the title "Emergency Interim Submission," which called for immediate action to curtail imports.[33] Since the several clothing associations (Apparel Manufacturers Associations of Quebec and Ontario, and the Manitoba Fashion Institute) normally submit separate briefs, and because the joint submission was strongly supported by the primary textile industry and the labour unions, the impact of this new industry-wide unity was considerable.

On November 8th, the Board presented its interim report to the IT&C Minister, concluding that "the categories of clothing listed below [there followed a list of 14 items, covering virtually all clothing types] are being imported in such quantities and in such conditions as to cause or threaten serious and immediate injury to production in Canada that would be difficult to repair."[34] The Board recommended on an interim basis, pending completion of the inquiry:

- that the Import Control List be amended to include the 14 clothing items listed;

- that the level of imports (of these goods) be limited in 1977 to the level of imports from all sources during 1975; [and]

- that, in those cases where limitations have already been implemented, imports in 1977 be limited to the level of imports in 1975, or to the level of the limitation, whichever is lower.[35]

(a) The Decision

Between November 9th and November 29th, meetings of the Interdepartmental Committee on Low Cost Imports and the Cabinet Committee on Economic Policy took place which determined the outcome of developments to that point.

(i) LCI Committee Meetings: Officials from the departments of Finance, External Affairs, Consumer and Corporate Affairs and CIDA were opposed to three aspects of the TCB recommendations: the broadness of product coverage of the proposed roll-back of imports to 1975 levels; the global nature of the roll-back; and the setting aside of existing voluntary export restraint agreements. Since the growth of import penetration in 1976 had ranged from 9 percent to 370 percent, it was claimed that not all of the 14 clothing items deserved the roll-back. The global scope of the roll-back was questioned on grounds of the wide divergence in the degree to which various countries were "responsible" for penetration, as well as the possibility that "non-responsible" trade partners might retaliate.[36] External Affairs was particularly concerned about the wisdom of the proposal to set aside existing VER's. Consumer and Corporate Affairs was especially worried that restraining imports to 1975 levels would be a heavy penalty for consumers. PCO and CIDA called for special consideration of the interests of the lower income developing countries.

Officials in the International Trade Relations Branch of IT&C either agreed with other departments' objections or were amenable to compromise. Mr. Chrétien's first major intervention during November, however, was to place strong pressure on his own officials:

> We in Finance and External might have got concessions from IT&C officials. But Chrétien had them under the gun saying "I want action and quick." After that the IT&C officials were unbending.[37]

(ii) The Cabinet Decides: How did Jean Chrétien persuade his Cabinet colleagues to accept the TCB recommendations in full? In spite of their officials' opposition to key aspects of the recommendations, the two other key Ministers, Don Jamieson of External and Donald MacDonald of Finance, succumbed to "a great selling job" by Chrétien.[38] The arguments which convinced Jamieson and MacDonald were the following:

- The controls were to be interim in nature, pending completion of the TCB inquiry.

- Unemployment was as high as the Government was low in the opinion polls.

- The TCB recommendations were the speediest route possible.

- The victory of the Parti Québécois in the Quebec election only a few days before had made the other ministers acutely sensitive to the "need" to use foreign economic policy to pursue national unity goals. In particular, the probable link between employment levels and the federalist cause was alluded to.

Donald MacDonald apparently needed more convincing than Jamieson, who had recently come from IT&C and no doubt vividly recalled the pressure he had been under from the Canadian textile industry and unions. As a Newfoundlander, moreover, the problem of regional unemployment was one he understood well. MacDonald, on other hand, was more inclined to be suspicious of protectionism. And Tony Abbott, the Minister of Consumer and Corporate Affairs, shared the same inclination, particularly as he was a past president of the Retail Council of Canada and had been involved in these import questions before in that capacity. It appears that the key factors in convincing MacDonald and Abbott were the Quebec election results and the acceptance of Chrétien, after Trudeau himself, as <u>the</u> spokesman for Quebec in Ottawa. It would have been difficult, even for powerful ministers like MacDonald and Jamieson, to challenge Chrétien's contentions about the link between unemployment and separatism in Quebec. And the one man who might have done so, the Prime Minister, did not intervene. Although Mr. Trudeau is known to have relatively liberal views on trade policy, it was not his custom to intervene regularly in the decisions of Cabinet Committees, and in the post November 15th atmosphere he was probably inclined to give the benefit of any doubt to his IT&C Minister. Jean Chrétien, the anti-separatist man of action, had won the crucial round two of the decision.

The Cabinet decision was announced by Chrétien on November 29th and became effective November 30th. In accordance with the "rational-political" model, ministers quickly developed a consensus based on their roles as party politicians and set aside those departmental and regional roles which had guided their behaviour on special import policy in the past. Concern for national unity overrode the fear of foreign retaliation which had usually militated against the use of the GATT Article 19 instrument. Bureaucrats in the various departments continued to hold divergent views on the Textile and Clothing Board's recommendations. Had the matter not been taken from their hands, they no doubt would have reached some sort of compromise different in many respects from the November 29th decision reached by their political masters. But with Cabinet temporarily united on the issue, the bureaucrats became largely irrelevant to this decision.

(iii) Postscript: Subsequent events indicate how ephemeral such cases of political consensus may be in the field of trade policy. Once the crisis aspects of the November decision had worn off, many elements of the "traditional" or "standard" governmental politics processes began to re-emerge. It would appear that follow-up decisions on clothing imports were once again made via a process of bargaining and compromise at both bureacratic and Cabinet Committee levels. Whereas the final report of the Textile and Clothing Board, issued in May 1977, recommended the negotiation of five year VER's with 21 exporting countries,[39] a compromise was reached whereby only the seven largest exporters would be subject to restraint and for only three years.

NOTES: Chapter Six

1. Canada has also placed many agricultural goods under quantitative restraint. The decision-making process for these products will not be covered here, as it is essentially separate from the process relating to manufactured imports.

2. Department of Industry, Trade and Commerce, Press Release No. 113-76, 29 November 1976.

3. See, for example, Caroline Pestieau, The Canadian Textile Policy: A Sectoral Trade Adjustment Strategy? (Montreal: C.D. Howe Research Institute, 1976); Caroline Pestieau, The Quebec Textile Industry in Canada (Montreal: C.D. Howe Research Institute, 1978); Department of Industry, Trade and Commerce, The Canadian Primary Textiles Industry, Sector Profile (Ottawa, 1978); Federal-Provincial Relations Office, The Textile Industry - A Canadian Challenge: A Report in the series Understanding Canada (Ottawa: Minister of Supply and Services Canada, 1979).

4. These and the following statistics are taken from Textile and Clothing Board, Clothing Inquiry: A Report to the Minister of Industry, Trade and Commerce (Ottawa, 29 May 1977).

5. Economic Council of Canada, Looking Outward: A New Trade Strategy for Canada (Ottawa, Information Canada, 1975) Appendix A, p. 196.

6. "Base levels" refer to the level of imports on which quotas are applied, usually related to some past import "performance." As for "growth rates," the Multifibre Agreement provides for growth of quota each year, usually by 6 percent, but sometimes higher for the lower-income developing countries.

7. For a fuller account of the agreement from a Canadian perspective, see Aaron J. Sarna, "Safeguards Against Market Disruption: The Canadian View,"

Journal of World Trade Law, Vol. 10, No. 4 (1976) pp. 355-370.

8. See Hon. Jean-Luc Pépin, Statement to the House of Commons on Textile Policy, 14 May 1970.

9. Montreal Gazette, 27 August 1976.

10. Montreal Gazette, 3 July 1976.

11. Confidential interviews.

12. Confidential interviews.

13. House of Commons, Standing Committee on External Affairs and National Defence, Minutes of Proceedings and Evidence Respecting International Development: Fifth Report to the House, 1st Session, 30th Parliament 1974-75-76, Issue 34 (April 8-13, 1976).

14. Senate of Canada, Proceedings of the Standing Committee on Banking, Trade and Commerce: Canadian Textile Problems: Report of the Committee, Issue 82 (April 7, 1976).

15. Confidential interview.

16. The present tense is frequently used in this section, since the committee is an ongoing mechanism which did not end in November 1976.

17. Confidential interview.

18. Confidential interview.

19. Confidential interview.

20. Confidential interview.

21. Confidential interview.

22. Donald W. Kelly, The Development of a New Textile Policy for Canada: A Case Study of Government-

Industry Relations in Canada, unpublished D.B.A. Thesis, Harvard University, 1974, pp. 272-273.

23. Textile and Clothing Board, Clothing Inquiry, op. cit., p. II-9.

24. Senate of Canada, Proceedings of the Standing Senate Committee on Banking, Trade and Commerce: Canadian Textile Problems, op. cit.

25. Montreal Gazette, 17 June 1976.

26. Ottawa Citizen, 18 June 1976.

27. Montreal Gazette, 31 July 1976.

28. Montreal Gazette, 10 November 1976.

29. Montreal Gazette, 10 November 1976.

30. C.B.C. Transcripts, Radio Noon, C.B.O. Ottawa, 7 January 1977.

31. Financial Times, 11 October 1976.

32. Montreal Gazette, 10 November 1976.

33. Highlights of the Emergency Interim Submission are reported in Textile and Clothing Board, Clothing Inquiry, op. cit., p. 6-1.

34. The interim report is published as Appendix 6 in Textile and Clothing Board, Clothing Inquiry, op. cit., pp. 6-4 to 6-5. The brackets are mine.

35. Ibid., p. 6-6.

36. The possibility was indeed real. Threatening retaliation, the United States demanded, and got, compensation in form of tariff reductions on some American exports to Canada.

37. Confidential interview.

38. Confidential interview.

39. Textile and Clothing Board, Clothing Inquiry, op. cit., pp. R-1 to R-11.

Chapter Seven

THE GENERAL PATTERNS OF TRADE DECISION-MAKING IN CANADA, 1968-1979

The simple four step model of the typical stages of government decision processes - initiation, review, bureaucratic resolution, political resolution - provides a convenient way of studying the basic structure and dynamic of trade decision-making in Canada during the Trudeau period.

(1) The Initiation Phase

Trade issues are most often initiated by either domestic interest groups or by foreign government actions and requests.[1] Less frequent, but still significant, initiatives come from the federal bureaucracy and political leadership. Earlier in this study, we presented in disaggregated form the views of major sectoral interest associations, as well as the links between these groups, provincial and parliamentary intermediaries, and their ultimate targets in the federal bureaucracy and Cabinet. At this stage, it is only necessary to present in summary form some conclusions about the aggregate strengths of Walter's "trade-biased" and "protection-biased" sectors.

The balance of power between inputs from the protection and trade-biased sectors in Canada appears to be influenced by three major factors:

- macro-economic conditions, where recessions favour protectionists and expansion favours the trade-biased sector;

- immediate political and electoral circumstances; and

- the context of decision-making where isolated decisions moderately favour protectionists and multilateral and bilateral negotiating rounds give an edge to the trade-biased sector, subject to the constraints listed above.

These conclusions are broad generalizations. It is quite possible, for example, that while a GATT round may lead to significant liberalization of Canada's tariff structure taken as a whole, certain sectors (e.g., textiles in the Kennedy Round) may succeed in largely avoiding losses of protection.

In addition, the above factors refer only to the relative strengths of inputs from the private sector. Outcomes in the form of decisions do not simply mirror these inputs: we have already observed how protectionist claims from the textile industry were "watered down" as they passed through the "standard" low cost import process, and how the success of demands for tariff increases is constrained by the GATT rules. A fuller understanding of trade outcomes therefore requires a look inside the review, bureaucratic and political resolution phases of the federal government decision-making process.

(2) The Review Phase

The review phase occurs most often when ministers, having received representations from domestic associations or foreign governments, or initiatives from within the government structure, direct their officials or advisory bodies to provide them with an informed judgement as to the issues at stake and the legal, political and economic implications of various options. Occasionally initiated by senior bureaucrats, especially on tariffs, most reviews are undertaken within the legally responsible department (IT&C, Finance or National Revenue) in consultation with other departments, most frequently External Affairs and Agriculture.

Because of this consultation, the lines of demarcation between the review phase and the bureaucratic resolution phase are often blurred. They are clearest where ministers exercise their second review option, that of reference to the specialized review mechanisms: the Tariff Board, the Anti-Dumping Tribunal or the Textile and Clothing Board. These references have tended to concern issues of long-standing controversy where full ventilation and an independent judgement seem both appropriate and politically expedient or because, as with textiles, clothing and dumping issues, the law requires it.

This review function is one of the major avenues of bureaucratic influence on trade decisions. The definition of the problem and issue at stake, as well as the setting out of options, will affect the way ministers make decisions. But ministers are generally far from pushovers. As a former Finance ADM and Canadian MTN negotiator notes about representations on the tariff:

> the very great number of such requests, and the great pressure of other Government business on the Minister, requires that he rely heavily on the advice of his officials as to which particular course of action he should adopt. But Ministers of Finance have, I think, always insisted that even such small and particular questions are put before them in sufficient detail as to enable them to make independent judgements, to bring to bear their own judgement of business and politics.[2]

(3) The Bureaucratic Resolution Phase

Technically, the bureaucratic resolution phase revolves around the preparation of a memorandum to Cabinet. According to quite formal criteria defined by the Privy Council Office, these memoranda are expected to contain several options of which one is recommended, to pass judgement on the impact of options on public opinion, federal-provincial and foreign relations, as well as to indicate the interdepartmental consultation patterns employed. In this section we shall examine the two basic patterns of interdepartmental participation in trade decision-making observable in the Trudeau era, the major patterns of alliance and cleavage between departments, and the typical means of resolving conflicts.

With respect to patterns of departmental participation in trade decisions, one point is clear at the outset. Few trade proposals ever reach Cabinet without the participation of the so-called "triumvirate" departments of Industry, Trade and Commerce, Finance, and External Affairs. Finance and IT&C are almost invariably involved in substantive issues. Although External's substantive input has increased in recent years, its role is still primarily one of coordination. Among secondary departments, Agriculture stands out as the only one which is inevitably an active participant, in its sphere, and which possesses the capability for sustained substantive input. For the others, such as Consumer and Corporate Affairs, Energy, Mines and Resources, CIDA, Regional Economic Expansion, Labour, National Revenue and the PCO, consultation and active participation have proven irregular.

We observe in the Trudeau years two broad patterns of bureaucratic actor participation: a formal committee approach, and an informal "big 3 plus 1" approach. The formal committee approach is the newer, being the result of reinforcements of the collegiality principle promoted by the Trudeau Government. By "formal committee" we mean, not that the committee is necessarily the result of a Cabinet directive (although most of them are), but that the committee's membership includes the full range of departments with some interest in the issues covered. Five major formal interdepartmental committees in the trade field existed during the Trudeau years, each corresponding to a particular statutory instrument or trade function. They were the Interdepartmental Committee on Low Cost Imports (for manufactured goods being considered for inclusion on the Import Control List), the Interdepartmental Committee on the Generalized System of Preferences, the Interdepartmental Committee on Economic Relations with Developing Countries (the trade portion of whose agenda has mainly concerned commodities), the Interdepartmental Committee on Trade and Industrial Policy (for industrial policy linkages to trade measures), and the Interdepartmental Committee on Commercial Policy (a would-be coordinating and "think-tank" body). Most have functioned officially at the ADM level, but in many cases de facto participation was at lower levels.

The second, and more traditional, pattern of departmental participation might be called the informal "big 3 plus 1" approach. Generally ad hoc affairs without a formal title, consultations often take place in very informal settings such as luncheons and telephone conversations between DMs, ADMs and Directors-General. Participation consists of Finance, IT&C and External plus one other department depending on the issue, although on occasion there may be two extra departments.

The policy coverage of this informal approach is very broad. The "big 3 plus 1" pattern applied in the 1968-79 period to all tariff matters (excepting the General Preferential Tariff since it was implemented in 1974), Import Control List questions other than "low cost" manufactured goods (i.e., primarily agricultural products), in practice though not in theory preparations for the Multilateral Trade Negotiations, some commodities issues, most coordinating functions, and most crisis situations (e.g., the August 1971 U.S. surcharge crisis).

What accounts for the existence of two different patterns of interdepartmental participation? In the pre-Trudeau heyday of the small community of "mandarins," the informal "3 plus 1" pattern was really the only one for trade decisions. It is the approach still preferred by senior officials in Finance and IT&C, while External has been keener on bringing other departments into the picture. Most of the DMs and ADMs in the triumvirate regard the formal committees as dreary, cumbersome mechanisms. They tend to believe that between themselves they can take account of all relevant considerations without complicating and slowing matters by having a broad range of departments involved in every issue. In fact, some formal committees died a gradual death as loci of real action because senior officials in the triumvirate tired of the process. Others, such as the Interdepartmental Committee on Commercial Policy, gradually held meetings less frequently and devolved onto lower officials.

Most of the formal committees had their origins in the PCO and PMO after pressure from secondary departments and ministers in the early 1970s. The policy-making philosophy of Prime Minister Trudeau and his closest advisors, emphasizing consultation and the creation of counterweight sources of advice, also contributed to the proliferation of such committees.[3]

Generally speaking, the formal committees are seen as painful necessities by Finance and IT&C, to be avoided if they can get away with it. The argument of budgetary secrecy has proven effective in restricting departmental participation on most tariff questions - and it is by no means always a specious argument. Some deft manoeuvres have also been used to limit the *de facto* input of secondary departments, such as varying the timing and distribution of agendas and demanding from secondary departments research inputs that they are known not to have the resources to produce. But all is not intrigue and jurisdictional jealousy. A major reason for the continuation of the informal "3 plus 1" pattern in spite of PCO preferences is the limited capability of most secondary departments to intervene effectively on all issues. Thus their participation in discussions may become a largely symbolic exercise without much contribution of substance, or they may take ideological positions based on inadequate homework, resulting in an atmosphere of conflict. This in turn tends to reinforce the Finance-IT&C desire to avoid them. When an informal "3 plus 1" pattern is extended to include another department, as happens occasionally with Consumer and Corporate Affairs, it is usually the result of strong interest in the issue by the minister of the extra department.

Patterns of alliance and cleavage in the bureaucratic resolution phase are fairly predictable, although they are not the same on every issue. In isolated instances of industry appeals for protection, External Affairs, Consumer and Corporate Affairs and CIDA officials provide consistent resistance to protectionism, the IT&C sector branches and most other "vertical constituency" departments are usually sympathetic to import controls, while Finance and the

International Trade Relations Branch of IT&C find themselves closer to the middle. We are speaking here of officials' preferences; actual behaviour depends greatly on both stated and inferred ministerial preferences. In preparations for multilateral trade negotiations, where the "fragmentation effect" is not paramount, cleavage patterns are somewhat modified. Constituency departments, such as Agriculture and the IT&C sector units, are forced to engage in internal reconciliation of the diverse interests of their strong and weak sectors, and the liberal IT&C sector divisions, representing strong manufacturing sectors, become involved. On the other hand, the "anti-protectionist" departments, such as External and CCA, retain this bias, and the more neutral Finance and the International Trade Relations Branch of IT&C tend to swing toward the liberalization side of the pendulum because of the possibility of trading opportunities for concessions in multilateral negotiations. These considerations help explain why most post-war Canadian trade liberalization has taken place in negotiating contexts. Of course ministers are capable of reversing this liberal bias of bureaucratic resolution in GATT rounds. However, since they are subject to many of these same considerations, they too tend to find liberalization more acceptable in international negotiations, although they are more cautious than bureaucrats.

Patterns of alliance and cleavage are also influenced by the two patterns of departmental participation described above. There is less conflict when participation is restricted to Finance, IT&C, External Affairs and one other department. As John Kirton has observed, "disagreements among the three central members, while frequent, are usually limited in degree and duration and confined to analytical points within well understood limits."[4] Thus, while different orientations between the big three have always existed, resolutions between them tend to take place within a spirit of problem-solving and willingness to compromise - an approximation of the "accommodation model" described earlier. This accommodation process is also encouraged by the fact that Finance and External Affairs, as well as the trade policy branch of IT&C, do not represent clear-cut constituencies. However, with the addition of a few other "vertical constituency" actors, as occurs in the formal interdepartmental committees, accommodation tends to break down and decision-making resembles more the "governmental politics" model. Of course, these scenarios both depend on ministerial preferences, since, as we observed in the case of clothing quotas in November 1976, ministers are capable of bypassing interdepartmental processes at the bureaucratic level when they perceive their basic political or party interests to be at stake.

How are differences between departmental officials resolved in the memorandum to Cabinet? To be sure, there are situations where resolution proves impossible and dissenting opinions are expressed in the memorandum. But certain norms of behaviour usually operate to limit the scope and duration of conflict. Probably the most important is the "lead department" principle. The department with legal responsibility for an issue has clear advantages in expressing its point of view. By tradition, the others do not press their case too far since they accept the basic legitimacy or primacy of the responsible or lead department. This holds true particularly for matters which are the domain of Finance, but probably less so for IT&C responsibilities since the two wings of IT&C are less likely to have a common position. Another principle is that agreement between the big three departments is very difficult to overcome, because these departments have the greater expertise and more prestigious ministers who are less likely to be challenged in Cabinet. In those few cases where a united big three are defeated, it is almost invariably accomplished in Cabinet, based on some other minister's invocation of political or regional considerations, or by prime ministerial intervention, rather than at the bureaucratic level. When actor participation is limited to the big three departments, resolution (which does not necessarily mean full agreement) is usually quicker than in committees having a large number of participants.

(4) The Political Resolution Phase

Because they simultaneously occupy departmental, regional-constituency and party roles, all entailing attention to trade issues, trade policy is a relatively high profile area of public affairs to most Cabinet ministers. Some ministers moreover have personal views and theories on the subject. While only a minority of ministers display an interest in international relations in general, the large majority are keenly sensitive to changes in the structure of protection, and are constantly made aware of such issues by pressure groups and provincial governments.

The "mandarin" theory notwithstanding, ministers in the end make the decisions. They are far from being the mouthpieces of their senior advisors. It is not uncommon for ministers to reject, or modify, the advice of officials. And if officials were not in the habit of "tailoring" their advice to the political realities, and anticipating the attitudes of ministers, there would doubtless be many more cases of rejection or modification.

While these statements would seem broadly valid for all administrations, the Trudeau Government encouraged participation in trade decisions by a larger number of ministers than was the case in the past. During the 1950s and 1960s, trade policy was the exclusive domain of Finance, Trade and Commerce, External Affairs and Agriculture. In concert with his desire to make Cabinet a truly collegial body, Mr. Trudeau reorganized the Cabinet committee system to permit a greater variety of ministers, who became more numerous with the creation of several new departments, to participate in each committee.[5] The new rules also allowed any minister, if so motivated, to attend a committee meeting of which he was not a member, and Cabinet committees were given real decision-making authority by making any committee decision final unless contested in full Cabinet.

Another purpose of the 1968 Cabinet reforms was to modify what Mr. Trudeau saw as an unhealthy balance of power between ministers and "mandarins." Like many others, he believed that a small coterie of deputy and assistant deputy ministers wielded extraordinary power over ministers by informally resolving issues among themselves and facing Cabinet with a united bureaucratic front.[6]

Trudeau sought to change this relationship by shifting the locus of most decisions from full Cabinet to smaller Cabinet Committees in which ministers could specialize and gain expertise, and by upgrading the resources of the PCO and PMO to provide Cabinet committees with alternative sources of advice. Ministers would no longer depend on any one DM or ADM. As Mitchell Sharp observed:

> The Trudeau approach to decision-making in the Cabinet has had many consequences. Perhaps one of its most significant has been to require ministers to become knowledgeable, even expert, to an extent that was not required of them in the past.... My impression is that for these reasons ministers are not as dependent upon their principal civil servant advisors for policy guidance as they were in earlier administrations and that interdepartmental committees, while they remain numerous, are not as significant in the decision-making process as they once were.[7]

In practice the Trudeau system entailed only a relative, not absolute, decline in the power of senior bureaucrats. Again, Mr. Sharp points out that "while ministers have become more expert, the problems have become more complex, and there remains plenty of scope for the exercise of the analytic powers and judgement of senior permanent advisors."[8] The trade policy influence of deputy ministers and other senior bureaucrats is undoubtedly greater in the more technically difficult and complex issues, in implementation activities, and in situations of rapid turnover of ministers or where a department is headed by an intellectually or politically weak or inexperienced minister.

The degree of personal involvement of ministers varies according to the question under discussion. Any issue on which Government caucus members or provincial authorities have lobbied almost always triggers the keen interest of ministers. Quantitative restrictions, and the more politically sensitive tariff questions, receive much greater and more regular ministerial input than the more numerous but largely routine tariff decisions. Because of the complexity of the Multilateral Trade Negotiations and the scarcity of time, ministers concentrated upon the analysis of packages of proposals prepared for them by bureaucrats rather than detailed consideration of each item.

But even in those situations where the personal involvement of ministers is quite low, it would be a mistake to conclude that they are unimportant in the policy process. Bureaucrats are continuously involved in a process of anticipating ministerial reactions, and trying to give advice which they feel will take account of their ministers' political, regional and departmental concerns. This advice is quite often at some variance with the personal preferences of the bureaucrats involved. Senior officials have generally proven quite astute as anticipators, and a large number of trade decisions are made with minimal personal involvement by ministers. Contrary to appearances, this fact does not imply that it is really the senior public servants who run the show.

This conclusion is somewhat different from those drawn by many analysts of other areas of Canadian public policy in recent years, particularly in the context of the accountability controversy.[9] These observers may well be right about the dominant role of senior bureaucrats in other fields. This study indicates, however, that because most trade issues are not particularly complex (in the technical or scientific sense) and are politically sensitive enough to be instinctively understandable by politicians, the direct or indirect role of ministers in trade decisions should not be minimized.

In the Trudeau decade, trade decisions were seldom made in full Cabinet. Most were considered in the

Cabinet committees on Economic Policy, Government Operations, and External Policy and Defence. Of these, the key committee was Economic Policy. Occasionally, broader issues, such as general industrial policy and the Economic Council's Looking Outward report, were discussed in Mr. Trudeau's committee of his most senior ministers, Priorities and Planning. The exact membership of the Economic Policy Committee was secret, but it certainly included the key economic departments plus other ministers based on the criterion of regional balance. Other ministers were entitled to attend if they wished. The Secretary of State for External Affairs, for example, was not a member after 1974, but this minister did attend numerous meetings concerned with trade and other international economic questions. The Cabinet also has a long tradition of *ad hoc* committees. In 1977 a special committee of 16 ministers on the Multilateral Trade Negotiations was set up under the chairmanship of Deputy Prime Minister Allan MacEachen. The large number of ministers on this committee, some of whose portfolios had nothing to do with trade policy, reflected the regional and political importance of the Geneva negotiations.

Patterns of ministerial participation in decision-making were influenced not only by the Trudeau application of the collegiality principle, which tended to expand participation, but also by the legal and informal hierarchy of ministers on trade issues that limited participation and influence. It is clear that the Ministers of Finance, IT&C and the Secretary of State for External Affairs are "more equal than the others" in making trade decisions. This is true not only by reason of their responsibilities, but also because these ministers are usually persons of great personal influence in Cabinet. Perhaps the majority of trade proposals between 1968 and 1979 went to Cabinet committees signed by all three. The second tier of influence is occupied by the "vertical-constituency" departments of which Agriculture remains paramount. Given his responsibilities, it is possible that the Minister of Consumer and Corporate Affairs might have played a consistently greater role but for the frequent turnover of incumbents of mostly low Cabinet standing. In addition, there exists a residual category of ministers who for personal or regional reasons may from time to time intervene. Theoretically, one might put Prime Ministers at the top of the hierarchy, but recent PMs have intervened little in trade policy.

Patterns of alliance and cleavage with the Cabinet on trade policy are less predictable than at the bureaucratic level, because of ministers' possession of multiple roles (department head, party politician, regional spokesman) and the possibility of role conflict. One statement, however, seems to apply to most ministers: they have tended to be more protectionist and have had less incentive to think about the long term than have their public service advisors. This seems less the result of attitudes and beliefs than of being more subject to political pressures from interest groups and the provinces. This pressure is less determinant of ministers' behaviour in multilateral bargaining situations than it is in the context of isolated industry representations, but even in the multilateral instance ministers seem more cautious about liberalization of barriers than do their officials.

Beyond this general principle, cleavage patterns in Cabinet reflect greatly the case-by-case choice of roles (or mix of roles) by ministers. We previously defined (in Chapter One) the "governmental politics" model as one where ministers bargain in Cabinet committee on the basis of their departmental or regional roles; the "rational-political" model as where shared party electoral interests predominate; and the "accommodation" model as where uncertainty or a desire to take account of multiple roles leads to a conscious accommodation of competing concerns. Our case study, as well as other evidence presented in this volume, indicates that during the Trudeau period there were examples of the operation of all of these models - for example, the governmental politics model in the "standard" low-cost import cases, the rational-political model in the extreme case of 29 November 1976, and the accommodation model in the formulation phase of the General Preferential Tariff (see next chapter). We also find that a lack of clear direction from ministers sometimes permits the operation of one model (for example governmental politics) at the bureaucratic discussion level and another model (often accommodation) at Cabinet level. In numerical terms, the governmental politics and accommodation models appear to be most frequent. Although the nature of this study does not permit quantitative precision, it is our impression that, at Cabinet level, accommodation is more frequent than governmental politics, partly because ministers are colleagues of the same party who can appreciate each others' concerns. It seems only somewhat more frequent at bureaucratic levels. Rational-political style shared images are much less frequent but do concern some very important decisions. We shall shortly, after reviewing the trade policy "subsystems," present several hypotheses about the conditions governing the applicability of each of these models.

The role of the Prime Minister merits further comment. Since Mackenzie King, Canadian Prime Ministers have intervened very little in trade policy. Diefenbaker did so more than either Trudeau or Pearson, but was known for intervention into the activity of many of his ministers.[10] Both Pearson and Trudeau, but especially Pearson, left trade policy to their Ministers of Finance and IT&C and accorded them full confidence. For example, beyond the signing the document, Mr. Pearson was hardly at all involved in the preparation of the Auto Pact, and the same applied to the Kennedy Round negotiations. All three of these Prime Ministers held trade views somewhat more liberal than their Cabinets, but were cognizant of the political dangers of conspicuous liberalization initiatives.

Mr. Trudeau's few interventions on trade questions were based on his philosophical beliefs and on issues whose diplomatic and economic content were blurred. The two areas of strongest Trudeau interest were international development - it was Trudeau who squashed a 1975 Cabinet committee decision to withdraw the General Preferential Tariff on scissors and who, along with Ivan Head, committed the country to UNCTAD's "integrated commodities program" - and the Third

Option, especially the "contractual link" with Europe. In most other trade decisions Mr. Trudeau's interventions were quite rare and tentative. The reason appears to lie in the always scarce resource of prime ministerial time, confidence in the relevant Ministers and Trudeau's relatively low interest in such matters. Even the response to such major trade situations as the 1971 surcharge crisis, and the Multilateral Trade Negotiations, was left entirely to ministers. Considered as a target of interest group lobbying, Prime Ministers are of course not very accessible. Perhaps the most important trade role of Trudeau and other Prime Ministers has been to enter into contact with other heads of government when action appears to require high level breaking of logjams. Most such cases have been at the request of ministers rather than prime ministerial initiatives.

NOTES: Chapter Seven

1. Although Canadians elected a new Conservative government on May 22, 1979, the present tense is frequently employed in this chapter to make points which we have no reason to believe will change under the Clark administration. Descriptions reflecting practices clearly unique to the former Liberal government are in the past tense. (To complicate things even more, the Conservatives were defeated in the House of Commons on December 31, 1979 and an election called for February 22, 1980.)

2. Rodney de C. Grey, personal communication.

3. G. Bruce Doern, "The Development of Policy Organizations in the Executive Arena," in G. Bruce Doern and Peter Aucoin (eds.), The Structures of Policy Making in Canada (Toronto: Macmillan, 1971).

4. John Kirton, The Conduct and Coordination of Canadian Government Decision-Making Towards the United States, unpublished doctoral dissertation, Johns Hopkins University, 1977, p. 74.

5. M.J.L. Kirby, H.V. Kroeker and W.R. Teschke, "The Impact of Public Policy-Making Structures and Processes in Canada," Canadian Public Administration, Vol. 21, No. 3 (Fall 1978) pp. 407-417.

6. Ibid., p. 409.

7. Mitchell Sharp, "Decision-Making in the Federal Cabinet," Canadian Public Administration, Vol. 19, No. 1 (Spring 1976) p. 6.

8. Ibid., p. 6.

9. The most recent, and best documented, discussions of this issue can be found in Colin Cambell and George J. Szablowski, The Superbureaucrats: Structure and Behavior in Central Agencies (Toronto: Macmillan, 1979); and in Royal Commission on Financial Management and Accountability [Lambert

Commission], Final Report (Ottawa: Minister of Supply and Services Canada, 1979).

10. See Peter C. Newman, Renegade in Power: The Diefenbaker Years (Toronto: McClelland and Stewart, 1963).

Chapter Eight

SELECTED TRADE POLICY SUBSYSTEMS

Having examined trade decision-making processes at a general level in the period 1968 to 1979, this chapter narrows in on a selection of policy "subsystems" in an attempt to achieve greater precision in our conclusions about the overall process. These subsystems are: unilateral tariff changes between, or unrelated to, GATT rounds; the Multilateral Trade Negotiations 1973-79; and the "technical" or depoliticized anti-dumping and countervailing duties processes. When added to our detailed look at special import policy in Chapter Six, comparative examination of these subfields should allow us to arrive at some useful statements about the workings of the models of decision-making described at the beginning of this study.

(1) Unilateral Tariff Changes

Tariff decisions are made in two ways: through international negotiations on the basis of reciprocity, and by unilateral action of the Canadian government. Unilateral decisions, on which this section focuses, are usually the result of either interest group representations or government objectives in macro-economic or balance of payments policy (e.g., in the annual budget). The General Preferential Tariff (GPT) may also be regarded as a unilateral tariff since, although it evolved out of international negotiations, it is not based on reciprocity between Canada and developing countries.

Tariff decisions may take several legal forms. At any moment there exists a schedule of duties passed by Parliament known as statutory rates. Parliament has delegated to Cabinet the right to make certain changes in these duties by order in council. Under the Customs Act and Customs Tariff, rates on products used as inputs into Canadian manufacturing and on chemicals and plastics may be temporarily reduced or eliminated (usually for one year) or restored to their statutory level; GPT rates may be restored to previous MFN rates on condition of injury to Canadian producers. All other changes, including changes in duties on end products, must be approved by a bill in the House of Commons.

In addition, there are "bound," as distinguished from "unbound" rates, i.e., those previously agreed to by Canada on a contractual basis in GATT rounds and bilateral negotiations. Today, the large majority of most favoured nation rates are bound. Although Canada can legally change bound duties, increases in such rates are subject to the compensation constraint of Article 28 of the GATT. For this reason, bound and MFN rates are in fact seldom increased. Thus when an interest group demands and gets a tariff increase, it is most often a matter of restoring a previously reduced rate to its statutory level or restoring to MFN a previous unilateral concession such as a GPT duty.

(a) The Unilateral Tariff Decision Sequence

Interest group appeals for increases or reduction of rates of duty are formally directed to the Minister of Finance. In practice, because of the large number of such claims, and the time constraints faced by the Minister, numerically the largest portion of representations go to the ADM for International Trade and Finance or to the Director of Tariffs Division. Those appeals which are addressed directly to the Minister are generally of most obvious political sensitivity, or are referred to him by other ministers and MPs.

Review of these appeals is in most cases (probably 95 percent) undertaken by the Tariffs Division of the Department of Finance. Although only a few such cases are sent to the Tariff Board, these usually involve complex and politically important issues. The Tariffs Division consults with other interest groups and with other departments, most frequently the line branches of IT&C and the Department of Agriculture for their sector expertise. Consultation with External

Affairs is in most cases largely pro forma and with Consumer and Corporate Affairs limited to cases where an increase in rates is being contemplated. The involvement of External and CCA, restricted by the inadequate resources of these departments to intervene effectively on the large number of "small" tariff cases, is substantial only in instances of diplomatic sensitivity like the GPT or where the consumer impact is evident. Thus while the informal "3 plus 1" pattern of departmental participation is basically characteristic of unilateral tariff changes, de facto participation is often even more restricted. Indeed, there have been cases where consultation did not extend beyond the Department of Finance. The only tariff slice whose participation patterns reflect the interdepartmental committee approach with the full range of departments is the General Preferential Tariff.

The large number of tariff appeals, and the demands on the Minister's time, make him quite dependent upon the analysis of his officials in Tariffs Division. Groups of 30 or so recommendations are often sent to the Minister for his approval, and it is sometimes the case that he was not aware that a review was being conducted until presented with a recommendation. Most of the time, officials' recommendations are accepted by the Minister. Where they are not, it is almost always for political reasons. This is not to imply that the Minister is merely a rubber stamp in tariff matters. On the whole, the senior tariff officials have been sufficiently astute politically to present recommendations which will be accepted, sometimes against their own preferences.

Tariff changes are also brought forward in the annual budget. Such actions generally derive from the government's anti-inflationary or employment-stimulation goals of the day, although satisfaction of narrower interest group claims is not unknown in budgets. While it has opened up somewhat in recent years, consultation with other departments is normally more restricted in the context of budgetary secrecy than in other situations. As Phidd and Doern observed:

> The work of the Branch [sic] was well illustrated in the 1973 budget which introduced a wide range of reductions in tariffs on agricultural products. The work was carried out through an interdepartmental liaison group including the departments of Industry, Trade and Commerce and Agriculture. A list of potential items was drawn up after the idea was initiated with the Deputy Minister of Finance.... In the early stages of policy development the proposals were discussed with the relevant departments. However, when the proposals reached an advanced stage they were discussed ... with the Minister of Agriculture but not with officials below.[1]

The decision of the Minister of Finance is normally taken to Cabinet Committee, which from 1968-79 was usually the Committee on Economic Policy. The large number of isolated cases, as well as the authority and prestige of the Minister of Finance, ensure that probably most individual tariff proposals are approved in committee with little opposition. Still, there are many examples of keen interest by non-responsible ministers. These correspond generally to cases of high profile consumer products, diplomatic sensitivity, regional sensitivity or cases where ministers have been subject to particular pressures from interest groups, provinces, and MPs in caucus. Some tariff questions, however, may never be considered in Cabinet Committee due to budgetary secrecy.

The bulk of interest group appeals for tariff increases are rejected these days, in large measure owing to the now solid international norm against tariff hikes, as well as the existence of industrial policy substitutes and alternative instruments of protection like quantitative restrictions, surtaxes and administrative barriers. In fact, during the Trudeau era there were numerically more instances of (temporary) unilateral tariff reductions than of increases. But this is probably due less to the success of anti-protectionist lobbyists than to the anti-inflationary efforts of the Trudeau government. On balance, the major bias in unilateral tariff decisions between GATT rounds would appear to lie close to the status quo, primarily because of the constraints on raising duties and because the traditional policy of reciprocity tends to reserve downward changes in statutory tariffs for multilateral negotiations.

(b) Subsystem Dynamic

The interaction of actors in unilateral tariff decisions corresponds primarily to the "accommodation model." While differences of view exist at both bureaucratic and ministerial levels, according to departmental interests and on regional grounds, the decision-making process has in most cases been characterized by conscious problem-solving as opposed to conflictual bargaining. A major reason appears to be an unwillingness on the part of Finance to subject tariffs to the kind of interdepartmental bargaining characteristic of IT&C's instrument, quantitative import controls. Finance usually consults, on its own terms, only with a restricted number of departments. The tradition of budgetary secrecy, the prestige of the Finance Minister and the reputation of the department as being beyond identification with particular interests, have allowed Finance largely to avoid PCO-inspired interdepartmental committees.

The thesis on which this study is based included a case history of the formulation and implementation of the Canadian General Preferential Tariff. It illustrates both the typical accommodation process of unilateral tariff decisions, and a major example of a breakdown of Finance's preferred technique of consultation as compared with shared decision-making.[2] Here, following, is a brief summary of that case study:

The General Preferential Tariff (GPT) came into effect in July 1974 and offered a 10-year program of tariff reductions on about 1,500 products for developing countries. The formulation phase of the GPT occurred between 1969, when Canada announced its intent to implement UNCTAD's recommendations on preferences for these nations, and May 1972, when Bill C-172, an

<u>Act to Amend the Customs Tariff</u>, was introduced in Parliament. The major decisions during this formulation phase concerned product coverage and the exceptions list. Four factors appeared to have encouraged an accommodation process.

(i) Departmental participation in decision-making was based on the "3 plus 1" principle, with Finance, External and IT&C the key actors, and Agriculture and Revenue Canada occasional participants. Other "vertical constituency" actors were little involved.

(ii) None of the participating departments took hard-line positions. Finance, the responsible department, was open to suggestions from the others, partly because it was not, in the secretive atmosphere of tariff formulation, under significant pressure from domestic interest groups, and partly because a 1969 understanding between UNCTAD and OECD allowed industrial countries to exclude the most sensitive products <u>ab initio</u> and to invoke safeguards in the event of future injury from imports. These same considerations encouraged flexibility on IT&C's part. External Affairs, the department most sympathetic to the program for reasons of Canada's international image, was also flexible. Its fear that the GPT, offered to all developing countries, might undermine the value of existing preferences to Commonwealth developing states discouraged it from advocating a "maximum" GPT package. This mix of concerns, and lack of adamant positions on the part of participating departments, favoured a problem-solving approach to GPT formulation.

(iii) The parameters of the decision process also encouraged moderation on the part of IT&C and External. The aim of the exercise was essentially to come up with a plan roughly as favourable to UNCTAD as those of other donors, most of whom had implemented their schemes in 1971, but at the same time no more dangerous to domestic interests. These parameters helped satisfy both External's preoccupation with Canada's image, and, through exceptions and safeguards, IT&C's concerns about the most sensitive industries.

(iv) The fact that GPT formulation involved a considerable slice of the Canadian tariff encouraged accommodation by allowing each department, whose aims were in any case limited and flexible, to achieve several of its most valued objectives. For example, Finance allowed inclusion of several additional agricultural products, which was not required by the OECD-UNCTAD understanding, to satisfy External in "exchange" for further exclusions of manufactures to please the industry line branch of IT&C.

After the new tariff was implemented in July 1974, a large number of appeals were received from Canadian businessmen urging withdrawal of GPT rates. About 30 were taken seriously, these concerning such products as sheet glass, work gloves, radios, TVs, leather clothing, rubber footwear, scissors and steel.[3] After several other departments voiced their desire for a share in these decisions, Finance set up an Interdepartmental Committee on the Generalized System of Preferences (GSP Committee) to consider appeals for withdrawal. This post-implementation phase of the GPT story, which lasted approximately two years, exhibited most of the characteristics of the "governmental politics" decision model.

First, the committee included the full range of interested departments: Finance, IT&C, External, CCA, CIDA, Labour and National Revenue. Each department came to "represent" its "constituency" or function, and thus the employment, consumer, business, diplomatic, development or labour dimensions of the issues.

Second, while this representational process should not be surprising insofar as the "vertical constituency" departments are concerned, two new factors arose to change the behaviour of the "big three" departments. The moderation that External Affairs had shown in the pre-1974 formulation phase was transformed into a firm resistance to withdrawals. This occurred because the "new international economic order" had become a significant factor in Canada's diplomacy, and was reinforced by the appointment of a new Secretary of State for External Affairs, Allan MacEachen, a man of abiding personal commitment to international development. At the same time, the proliferation of interest group lobbying after the introduction of the new tariff put pressure on IT&C, and to a lesser degree Finance, to take harder, protectionist stances.

Third, governmental politics were also encouraged by the fact that, in contrast to the earlier phase, decision-making was conducted on a product-by-product basis. Departments thus took less flexible positions than during the formulation phase, when they had believed that what was "lost" in one area of the package could be compensated for elsewhere.

The results of the GSP Committee process were mixed: in a few cases the protectionists or liberals achieved more or less complete victory. In most cases a form of compromise was reached, but sometimes only after the issues had reached Cabinet level.[4] By late 1976, it appeared that Finance was tiring of the atmosphere of conflict in the Committee. The group was seldom convened after 1976, and Finance reverted to informal consultations with a smaller number of departments.

To summarize, the unilateral tariff decision process during the period 1968 to 1979 was most frequently characterized by "accommodation politics" based on "3 plus 1" patterns of participation. Only in the post-1974 phase of the GPT were "governmental politics" frequently observed. The other decision models applied very rarely. However, one clear example of the "individual model" was found. In the case of quality scissors, Prime Minister Trudeau, largely on the basis of his personal commitment to international development, overturned a Cabinet committee decision which would have withdrawn the GPT duty from this product.[5] The "political model" was also infrequently observed. Its requisites, shared political values and situations of crisis, were uncommon, in large part because sudden surges in imports are usually dealt with through other instruments, such as GATT Article 19 action and surtaxes.

(2) The Multilateral Trade Negotiations, 1973-79

After six years of analysis, posturing and, finally, serious bargaining in Geneva, the agreements achieved in the Tokyo Round of GATT negotiations were announced in April 1979. These Multilateral Trade Negotiations (MTN) were pursued through seven negotiating groups, on tariffs, non-tariff measures, agriculture, safeguards, sectors, tropical products and "framework" issues (such as dispute settlement and special treatment of developing countries). The procedure for tariff reduction introduced by the Kennedy Round, that of agreeing on an across-the-board or linear tariff cutting "formula," and then negotiating over exceptions lists, was also followed in the Tokyo Round.

The concluding agreements produced new tariff schedules which lowered duties on industrial products by an average of 40 percent. Some products especially sensitive to each country were reduced by lesser amounts, and in a few cases duties were eliminated entirely. Agricultural tariffs were not negotiated on a formula basis, and consequently reductions in this sector were smaller than the industrial average. In addition, seven codes of behaviour respecting non-tariff measures were concluded, on subsidies and countervailing duties, revisions to the Anti-Dumping Code, technical barriers to trade, government procurement, import licencing, customs valuation, and an understanding on dispute settlement and legalization of special treatment in favour of developing countries. No agreement was reached on the difficult, essentially North-South, issue of interpretation of the GATT Article 19 safeguards clause.

Like most countries, Canada achieved substantial benefits from the negotiations while making significant concessions in return.[6] Gains included reduced duties on many key exports, most notably inorganic chemicals and plastics, some fish and farm products, alcoholic beverages and forest products, as well as the codes on government procurement and subsidies and countervailing duties (which will limit future American use of countervailing duties). Not all of these codes or tariff reductions went as far as Canada had hoped - the procurement code, for example, does not apply to telecommunications equipment, an area of solid Canadian industrial performance - but an attractive deal was reached with the United States which will see about 80 percent of Canadian exports enter the American market duty-free. Canada's major concessions to trade partners included abandonment, subject to certain safeguard conditions, of the "fair market value" system of customs valuation in favour of a regime based on export prices, modifications to the industrial machinery remissions program, liberalization of federal procurement policies, as well as the across-the-board tariff cuts.

(a) Organization of Canadian Participation in the MTN

Soon after the Tokyo Round was launched in 1973, an institutional structure, periodically adapted to meet changing circumstances, was set up to prepare Canadian participation in the negotiations. As in the Kennedy Round, a Canadian Trade and Tariffs Committee (CTTC) was established as a channel for communication of interest group views to the federal government. Over 400 briefs were submitted and a large number of meetings with private associations held. The CTTC was in fact an interdepartmental committee chaired by Mr. Louis Couillard, a former chairman of the Tariff Board, and was composed of the departments of Finance, IT&C, External, Consumer and Corporate Affairs, Agriculture, Energy, Mines and Resources, Labour, Regional Economic Expansion, Revenue, the PCO and the head of the Canadian delegation in Geneva. Businessmen did not limit their lobbying efforts to the Committee. Frequent direct contact with ministers and senior officials was made. A second information-gathering mechanism (though it was somewhat more than this) was the Federal-Provincial Committee of Deputy Ministers of Industry. The creation of this committee in 1975 was a compromise with the provinces, several of which had wished actual representation on the Canadian delegation in Geneva.

Information gathered by the CTTC was fed to the departments which performed the real review function in the preparation of tariff offer and request lists, and positions on non-tariff issues. Most of this industry-by-industry analysis was done by Finance and IT&C, both of which equipped themselves with special internal instruments - the "mini-prep com" in IT&C and the office of the General Director [sic] of International Trade and Finance Branch at Finance. An informal division of labour on tariff questions saw IT&C concentrate on the export interest or request lists and Finance on exceptions lists. Each performed the lead role on non-tariff questions falling under its responsibility - procurement and import licencing in IT&C and customs, anti-dumping and countervail policy in Finance. Substantial analytic work in their spheres was also performed by the Departments of Agriculture and Fisheries. External Affairs' contribution stressed non-tariff barriers codes, where its expertise in international law could be brought to bear, and the ticklish problem of trade relations between developed and developing countries.

Reports and position papers arising out of the review function were taken up in an interdepartmental committee, whose role was to forward recommendations to Cabinet for decision and instructions to the negotiating team in Geneva headed by Mr. Rodney de C. Grey, a former Finance ADM. Until 1977 this committee was called the Trade Negotiations Coordinating Committee (TNCC), and was chaired by the Undersecretary of State for External Affairs, largely as a compromise between the competing wishes of Finance, which had chaired the equivalent committee during the Kennedy Round, and IT&C, which alluded to the _Industry, Trade and Commerce Act_ to bolster its claim to leadership. The TNCC had as members the full range of departments, but in fact the plenary committee seldom met. The significant members were External, Finance and IT&C, operating according to the "3 plus 1" principle whereby other departments were consulted according to the subject matter at hand (e.g., Supply and Services on procurement issues).

By the Spring of 1977 the government came to believe that a tighter, more centralized organizational structure would be needed to manage Canada's responses in view of the accelerating pace of the negotiations. A special, or ad hoc Cabinet Committee of 16 Ministers, balanced according to departmental and regional criteria, was set up under the chairmanship of Allan MacEachen, Deputy Prime Minister. A new Office of the Canadian Coordinator for the Multilateral Trade Negotiations was established under J.H. "Jake" Warren, former Ambassador to the United States and Deputy Minister of IT&C, to better coordinate the activities of government departments and consultations with the private sector and provinces. Warren reported directly to Mr. MacEachen, and his office became a secretariat serving the ad hoc Cabinet Committee. Warren was also appointed chairman of the Federal-Provincial Committee of Deputy Ministers. A new interdepartmental committee was set up at the same time, comprising Mr. Warren (as chairman), the head of the Trade and Tariffs Committee, Assistant Deputy Ministers in Finance, IT&C and External, and occasional additional participants. The role of this committee remained essentially the same as the previous External-led Trade Negotiations Coordinating Committee - to forward recommendations to the MacEachen Committee for approval. It was reported by interviewees to have been more efficient though, given the centralization of coordination activities in the office of a full-time Coordinator for the negotiations.

Mr. Warren's office quickly became the primary means of linking the Cabinet Committee, the federal and provincial bureaucracies, private sector associations (Warren having taken over much of the liaison with private groups from the Trade and Tariffs Committee in the concluding phase of the MTN) and the delegation in Geneva. One observer has argued that Warren's office largely appropriated the Department of Finance's lead role in tariff decision-making.[7] This research, however, would question that conclusion. It is doubtful that Warren's small office of four or five people could compete with Finance and IT&C in expertise and analytic capability. It is also questionable that the Minister of Finance would agree to relinquish to a senior bureaucrat such important matters as his authority over tariffs, the Customs Act, and anti-dumping and countervailing duty policy.

But it is evident that Mr. Warren, who is highly expert on trade matters, became a very influential man in terms of substantive decisions as well as coordination of actors.

(b) Subsystem Process

The Ottawa decision-making process for the MTN exhibited, according to issues, a mixture of value concensus and differing perspectives between the major actors. The first phase of the process, lasting until about mid-1976, was primarily concerned with goal-setting activities, including preparation of a tariff and non-tariff request list (i.e., requests for lowering of trade partners' barriers to Canadian exports) and the Canadian proposal for a "sector approach" to the negotiations, aiming at comprehensive liberalization in a few key sectors of major export interest to Canada.

The "rational-substantive" decision model provides the best explanation for this early phase of the process. A broad concensus existed in Ottawa and the provinces on the desirability of agreements that would reduce foreign barriers to the upgrading of Canadian resources prior to export, particularly in the non-ferrous metals and forestry industries stressed in the "sector approach." All concerned could support requests that the Europeans liberalize their procurement regulations and that the Americans bring their countervailing duty legislation into line with international practice. This goal-setting phase was of course the easiest part of the MTN decision-making process since it is in the nature of initial positions at international negotiations to request an "ideal" package. To be sure, the Department of Finance was less enthusiastic than IT&C and External Affairs about the manner in which Canada's sector approach was articulated, believing it lacked "saleability." IT&C had hoped that similar American interests on several sector issues would give the proposal more clout with the EEC and Japan. Finance proved correct, however, when Europe and Japan rejected the sector approach outright and the U.S. came to regard it as a last resort technique.[8] But differences of opinion between Finance and IT&C on this issue were primarily tactical in nature. A broad concensus existed between departments and ministers as to export objectives, especially the upgrading of resources.

The next phase of the negotiations was the decision whether to accept the so-called Swiss formula for tariff reductions. This plan envisaged a 40 percent linear formula or basic cut, with negotiations over exceptions lists and a measure of "harmonization" to cut high tariffs more than low ones. During the Kennedy Round, Canada had been able to obtain a waiver on implementing a linear formula, successfully arguing that a linear cut on Canada's generally high tariffs would not be compensated by a similar percentage cut on its trade partners' nominally low but still effective rates of protection, particularly against processing of Canadian raw materials.

In spite of considerable opposition from the Canadian Manufacturers' Association and some provincial governments, Canada's agreement to the Swiss formula was announced in January 1978. It was accompanied by insistance that assent was conditional on deeper-than-formula cuts in resource-based sectors. This acceptance was pressed on Ottawa by its trade partners. Both Europeans and Americans, believing that Canada had gained more than it conceded in the Kennedy Round, had made it clear that Canada would have to participate in the negotiations as a formula country if any of its export objectives were to be achieved.

Having accepted, with some reserve, the formula, action shifted to the detailed negotiations with the United States, Japan and Europe over tariff lists and non-tariff issues. As these bilateral offers and counter-offers proceeded, disagreements arose within Ottawa between departments and ministers on a number of individual issues. But the dynamic of the decision process did not appear to accord with the "governmental politics" bargaining model. Rather, the "accommodation" or problem-solving model seems the better explanation. Several reasons may be offered:

First, on most issues, ministers and departments did not take hard-line, stereotyped stances. They were not on the whole protectionist or liberal across-the-board; positions varied according to individual products and non-tariff issues. The behaviour of the Minister of Finance is illustrative: Mr. Chrétien was quite protectionist on certain issues sensitive to Quebec, such as textiles and footwear, but was not noticeably protectionist on most other products. Most ministers displayed a similar tendency to have strong opinions about a small number of pet issues, but were open-minded on most questions.

Second, the nature of multilateral negotiations is such as to favour accommodation rather than conflict and bargaining in Ottawa decision-making. Because virtually the whole economy is subject to discussion, and the aim of the exercise is to achieve a balance of export opportunities against increased import competition, vertical constituency departments are forced to take a broader view than in isolated, one-product trade decisions. Departments like Agriculture, Fisheries and IT&C must attempt to reconcile the competing interests of their export-oriented and import-sensitive sectors, and their positions become relatively flexible. Ministers are encouraged to pursue a mix of priorities among their departmental and regional roles. Mr. Horner, for example, a free trader by inclination, became a "fair trader" in the MTN, and the Consumer and Corporate Affairs Minister recognized that political and regional sensitivities must be given equal weight to the imperatives of his Department's mission. In fact, a Cabinet decision early in the negotiating phase gave some formality to the "accommodation" or problem-solving process by directing bureaucrats to develop for consideration by Cabinet a balanced package reflecting the conflicting pressures.

Third, the high stakes involved in the MTN and the fact of negotiations with more powerful trade partners probably encouraged a form of "patriotism" which favoured accommodation rather than the expression of narrow concerns.

In summary, the "rational-substantive" decision model characterized the earliest and easiest phase of the negotiations. Most subsequent decisions appear in conformity with the "accommodation model." However, to these conclusions must be added the caveat that domestic-based decision models can offer only partial explanations of outcomes in an international bargaining context. A more thorough analysis would require examination of other governments' internal processes, and the dynamics of bargaining between states. These fall beyond the scope of this study.

(3) The Anti-Dumping and Countervailing Duty Subsystems

As a result of successful negotiation of international agreements under the GATT, Canada has established domestic legislation and regulations sufficiently precise to essentially remove anti-dumping and countervailing duties decisions from the political process. The *Anti-Dumping Act* was passed by Parliament in 1969 in response to a GATT Anti-Dumping Code negotiated during the Kennedy Round, while the Countervailing Duty Regulations were approved by Cabinet in March 1977 in anticipation of a forthcoming Subsidy/Countervailing Duties Agreement in the Multilateral Trade Negotiations.[9]

The structure of decision-making for anti-dumping action was described in Chapter Four of this study. The new countervailing duty regulations outline a similar procedure, by which the Deputy Minister of National Revenue initiates an investigation which may or may not yield a "preliminary determination of subsidization," and the Anti-Dumping Tribunal determines whether the subsidization is injurious to Canadian firms. Cabinet must accept the Tribunal's findings if these reject the imposition of countervailing duties, although it need not apply duties in the event that injury is found.

The anti-dumping and countervailing duty rules have essentially depoliticized decision-making by organizing each process into a series of technical and administrative steps undertaken by National Revenue and the independent Tribunal. The evidence so far indicates that both processes are virtually immune from political manipulation. In the first year of the Countervailing Duty Regulations, for example, about a dozen complaints of subsidization were filed by Canadian producers, but in no case did the Deputy Minister of National Revenue see sufficient prima facea evidence of subsidization even to warrant a formal investigation. In the case of cheese products, the regulations enabled the Deputy Minister to resist pressure from the Minister of Agriculture to push through countervail protection.[10]

These subsystems provide the purest examples found in this research of the workings of the "rational-substantive" model of decision-making. The legislation and regulations prescribe for decision-makers quite precise "objectives" which allow them to deduce conclusions in almost syllogistic fashion. Problems which exist in the two subsystems are more technical and definitional (e.g., just what constitutes injury?) than political in nature. In this sense, the anti-dumping and countervailing systems might be seen as models for future shifts of trade policy instruments away from the realm of politics into the domain of administration.

NOTES: Chapter Eight

1. Richard W. Phidd and G. Bruce Doern, The Politics and Management of Canadian Economic Policy (Toronto: Macmillan, 1978) pp. 241-242.

2. The full case study is contained in Chapter Eight of David R. Protheroe, Making Trade Decisions in Canada, unpublished M.A. Thesis, Carleton University, 1979.

3. Confidential interviews and Globe and Mail, 28 June 1974.

4. See David R. Protheroe, Making Trade Decisions in Canada, op. cit., pp. 213-227 for details.

5. Confidential interveiw.

6. Opinions differ as to how Canada fared in the negotiations. Most of the MTN principals have estimated a rough balance of Canadian gains and losses. See, for example, Announcement of the Honourable Allan J. MacEachen, Deputy Prime Minister, on the Results of the Multilateral Trade Negotiations in Geneva, Press Release, Office of the Deputy Prime Minister, Ottawa, 12 April 1979. On the other hand, Canada's chief negotiator in Geneva, Mr. Rodney de C. Grey, who retired from public service at the conclusion of the MTN, has dissented from the majority view and presented a more pessimistic account of Canadian trade opportunities after the Tokyo Round. See Rodney de C. Grey, "How to Deal with U.S. Import Law," Financial Post, 24 November 1979, p. 1; and Ronald Anderson, "Lack of Canadian Clout Reflects Trade Realities," Globe and Mail, 31 August 1979, p. B2.

7. Gilbert R. Winham, "Bureaucratic Politics and Canadian Trade Negotiations," International Journal, Vol. 34, No. 1 (Winter 1978-79) pp. 64-89.

8. Caroline Pestieau, The Sector Approach to Trade Negotiations: Canadian and U.S. Interests (Montreal: C.D. Howe Research Institute, 1976) p. 11.

9. Canada Gazette, Part II, S.O.R. 77-271, 13 April 1977.

10. Confidential interviews.

CHAPTER NINE

SUMMARY AND CONCLUSIONS

It is clear from this study that no single decision model can explain all trade decisions taken between 1968 and 1979. Nonetheless, the wide range of decisions surveyed here does permit a few conclusions about the relative applicability of these models and the conditions determining their relevance.

A generous measure of support has been found for Robert Presthus' "elite accommodation" theory and his portrayal of political life in Canada as essentially distinguished by concensus on many basic values (on means if not ends), stability, and incremental change.[1] The accommodation model, which acknowledges disagreements between policy-makers but sees decisions arising out of a cooperative problem-solving process where actors consciously seek to reconcile a mix of priorities and interests, appears to have characterized the bulk of trade decisions taken by the Trudeau Government, in particular most tariff decisions and the multi-issue GATT negotiations. Some general features of the Canadian political system identified by Presthus as promoting elite accommodation - such as Cabinet solidarity and a fundamental concensus among government and private sector elites on the proper means of pursuing group objectives - are also applicable to the trade decision process. In addition, accommodation is associated with certain aspects of the trade decision-making system itself: the "3 plus 1" departmental participation pattern; instruments for which the Department of Finance, an horizontal and central agency portfolio not bound to particular interests, is responsible; and multilateral negotiating contexts and even some unilateral situations like the formulation phase of the General Preferential Tariff, where a broad range of items on the agenda encourages non-conflictual give-and-take among officials and ministers.

If the "accommodation" model may be regarded as the norm, it is plainly not always applicable. The representational or "governmental politics" model appears to account best for issues of unusual political and economic controversy, of which the "standard" low-cost import process and the post-implementation phase of the GPT were prime examples. These questions had in common a wide variety of considerations - employment versus consumer interests, discordant regional sensitivities, the economic efficiency question, domestic priorities versus international obligations, and the moral-diplomatic international development issue - about which all or most participants inside and outside government held firm views. But no single interest or dimension was perceived as overwhelmingly compelling. When this is the case, events encourage action in accordance with the "rational-political" model. Governmental politics also seem related to situations where departmental participation expands beyond the "3 plus 1" pattern, and where the legally responsible minister is relatively open-minded.

The remaining models were applicable less frequently. We encountered only one clear instance of the "individual" model (the scissors case), although there may well have been others unearthed by this research. The collegial emphasis of the Trudeau Government probably accounts for the rare sightings of this model. It does, however, seem a useful supplementary model since it directs attention to the important, if not determinant, role of individuals in particular decisions (e.g., Mr. Chrétien's role in the November 1976 clothing decision).

Concensus on trade objectives is sufficiently rare as to greatly limit the applicability of the "rational-substantive" model. During the 1968-79 period it was characteristic only of the depoliticized anti-dumping and countervailing duties subsystems, and the easy early MTN decisions. Similarly, concensus about political objectives on trade issues was infrequent, but occurred occasionally in crisis situations, such as the clothing controversy of November 1976, when the party interests of ministers came to overshadow their regional and departmental roles.

Care should be taken, however, in extrapolating from our conclusions on the relative applicability of decision models to other areas of public policy. It may well be that Ottawa decisions on program budgets, for instance, reflect the "governmental politics" model to a greater degree than the trade decision-making process, which rotates largely around legislation and orders in council. This study has far from exhausted the possibilities of comparative testing of decision models.

The models employed here provide alternative ways of explaining how government actors behave and interact. A fuller explanation of decisions invites consideration of some of the factors influencing "who wins" (and who loses) in these interactions, particularly those described by the "accommodation" and "governmental politics" models (the question being moot in the individual and shared value models). This study suggests seven sources of bureaucratic and ministerial influence over outcomes. These are listed in rough order of precedence.

First, ministerial and departmental power is strongly connected to the legal responsibility of portfolios for trade instruments and related issues, in large measure because this authority correlates to expertise. The most important category is jurisdiction over a direct trade instrument (e.g., Finance and tariffs), but control over related activities (e.g., international agreements for External, food marketing programs for Agriculture) is also a significant avenue of influence. Absence of legal authority entails a sharp drop in influence, even when the department has a recognized interest in the issue (e.g., CIDA and trade relations with developing countries). Second, the prestige, talents, regional power base and experience of individual ministers account for much of the differential influence of departments, and may sometimes allow departments to make up for their lack of legal authority. Third, the perceptions of other actors of a deparment's objectivity on trade issues affects its impact. Virtually all secondary departments, as well as the sector divisions of IT&C, are regarded as having an "axe to grind," and this limits their influence. By contrast, Finance, External Affairs and to some extent the International Trade Relations Branch of IT&C are seen as less attached to particular interests, and this makes other actors more receptive to their arguments. Fourth, the political weight of a vertical department's client groups is relevant. The political strength of farmers is well out of proportion to their numbers, and contrasts with the weakness of CIDA's international constituency and CCA's ill-organized consumers. In a similar way, departments and ministers whose stances are supported by Government caucus MPs and provinces are greatly strengthened. Fifth, prevailing economic and social conditions will cause variations in the degree to which a department's or a minister's bias is heeded. Sixth, divisions within a department or an alliance of departments can entail loss of influence, as was occasionally noticed in the case of IT&C. Finally, bargaining skills may influence outcomes, particularly between the big three departments. But even tactical brilliance on the part of secondary departments is unlikely to overcome frequently the weight of legal responsibility, expertise and ministerial prestige.

Based on these factors, it is possible to suggest a hierarchy of departments in terms of their influence on trade policy during the Trudeau years. The Department of Finance stood at the top of this pyramid, by reason of its resources of statutory responsibility, expertise, perceived objectivity, prestigious Minister, relative lack of internal division, and respected bargaining skill. In its own fields of responsibility, Finance's weight was overwhelming. Less expected, on issues outside its legal jurisdiction Finance emerged as the virtual equal of IT&C due to the latter's lesser cohesion, perceived lesser objectivity, and somewhat lower status Minister. IT&C, of course, normally occupied the second position in this hierarchy. On agricultural issues, however, Agriculture took over second place, based on its expertise and the political weight of its client group. Third spot normally belonged to External Affairs, due to its high status Minister and lack of particularist ties (rather than expertise), but External Affairs slipped to fourth place on agricultural questions. The lowest rungs were occupied by the protectionist and pro-trade liberalization vertical constituency departments. It is difficult to say whether one group was lower than the other across-the-board. Their impact depended greatly on the specific issues, current economic and social conditions, and the individual qualities of their ministers.

NOTES: Chapter Nine

1. Robert Presthus, *Elite Accommodation in Canadian Politics* (Toronto: Macmillan, 1973).

Institute for Research on Public Policy

PUBLICATIONS AVAILABLE*

January, 1980

BOOKS

Leroy O. Stone Claude Marceau	Canadian Population Trends and Public Policy Through the 1980's. 1977 $4.00
Raymond Breton	The Canadian Condition: A Guide to Research in Public Policy. 1977 (No Charge)
Raymond Breton	Une orientation de la recherche politique dans le contexte canadian. 1978 (No Charge)
J.W. Rowley & W.T. Stanbury, eds.	Competition Policy in Canada: Stage II, Bill C-13. 1978 $12.95
C.F. Smart & W.T. Stanbury, eds.	Studies on Crisis Management. 1978 $9.95
W.T. Stanbury, ed.	Studies on Regulation in Canada. 1978 $9.95
Michael Hudson	Canada in the New Monetary Order - Borrow? Devalue? Restructure! 1978 $6.95

* Order Address: Institute for Research on Public Policy
P.O. Box 9300, Station "A"
TORONTO, Ontario
M5W 2C7

W.A.W. Neilson & J.C. MacPherson, eds. — *The Legislative Process in Canada: The Need for Reform*. 1978 $12.95

David K. Foot, ed. — *Public Employment and Compensation in Canada: Myths and Realities*. 1978 $10.95

W.E. Cundiff & Mado Reid, eds. — *Issues in Canada/U.S. Transborder Computer Data Flows*. 1979 $6.50

G.B. Reschenthaler & B. Roberts, eds. — *Perspectives on Canadian Airline Regulation*. 1979 $13.50

P.K. Gorecki & W.T. Stanbury, eds — *Perspectives on the Royal Commission on Corporate Concentration*. 1979 $15.95

David K. Foot — *Public Employment in Canada: A Statistical Series*. 1979 $15.00

Meyer W. Bucovetsky, ed. — *Studies on Public Employment and Compensation*. 1979 $14.95

Richard French & André Béliveau — *The RCMP and the Management of National Security*. 1979 $6.95

Richard French & André Béliveau — *La GRC et la Gestion de la Sécurité nationale*. 1979 $7.95

Leroy O. Stone & Michael J. MacLean — *Future Income Prospects for Canada's Senior Citizens*. 1979 $7.95

Douglas G. Hartle — *Public Policy Decision-Making and Regulation*. 1979 $12.95

Richard Bird (in collaboration with Bucovetsky & Foot)	*The Growth of Public Employment in Canada*. 1979 $12.95
G. Bruce Doern & Allan M. Maslove, eds.	*The Public Evaluation of Government Spending*. 1979 $10.95
Richard Price, ed.	*The Spirit of the Alberta Indian Treaties*. 1979 $8.95
Peter N. Nemetz, ed.	*Energy Policy: The Global Challenge*. 1979 $16.95
Richard J. Schultz	*Federalism and the Regulatory Process*. 1979 $1.50
Lionel D. Feldman & Katherine A. Graham	*Bargaining for Cities, Municipalities and Intergovernmental Relations: An Assessment*. 1979 $10.95
Elliott J. Feldman & Neil Nevitte, eds.	*The Future of North America: Canada, the United States, and Quebec Nationalism*. 1979 $7.95
Maximo Halty-Carrere	*Technological Development Strategies for Developing Countries*. 1979 $12.95
G.B. Reschenthaler	*Occupational Health and Safety in Canada: The Economics and Three Case Studies*. 1979 $5.00

OCCASIONAL PAPERS ($3.00 per copy)

W.E. Cundiff (No. 1)	Nodule Stock? Seabed Mining and the Future of the Canadian Nickel Industry. 1978
IRPP/Brookings (No. 2)	Conference on Canadian-U.S. Economic Relations. 1978
Robert A. Russell (No. 3)	The Electronic Briefcase: The Office of the Future. 1978
C.C. Gotlieb (No. 4)	Computers in the Home. 1978
Raymond Breton & Gail Grant Akian (No. 5)	Urban Institutions and People of Indian Ancestry. 1978
K.A. Hay (No. 6)	Friends or Acquaintances? Canada as a Resource Supplier to the Japanese Economy. 1978
T. Atkinson (No. 7)	Trends in Life Satisfaction. 1979
M. McLean (No. 8)	The Impact of the Microelectronics Industry on the Structure of the Canadian Economy. 1979
Fred Thompson & W.T. Stanbury (No. 9)	The Political Economy of Interest Groups in the Legislative Process in Canada. 1979
Gordon B. Thompson (No. 10)	Memo from Mercury: Information Technology is Different. 1979

Pierre Sormany (No. 11) — Les Micro-Esclaves Vers Une Bio-Industrie Canadienne. 1979

WORKING PAPERS (No charge)

W.E. Cundiff (No. 1) — Issues in Canada/U.S. Transborder Computer Data Flows. 1978 (Out of print; in IRPP book of same title.)

John Cornwall (No.2) — Industrial Investment and Canadian Economic Growth: Some Scenarios for the Eighties. 1978

Russell Wilkins — L'éspérance de vie par quartier à Montréal, 1976: un indicateur social pour la planification. 1979